T0286675

Science v. Story

Science v. Story

NARRATIVE STRATEGIES FOR SCIENCE COMMUNICATORS

Emma Frances Bloomfield

UNIVERSITY OF CALIFORNIA PRESS

University of California Press
Oakland, California

Library of Congress Cataloging-in-Publication Data

Names: Bloomfield, Emma Frances, author.
Title: Science v. story : narrative strategies for science communicators / Emma Frances Bloomfield.
Description: Oakland, California : University of California Press, [2024] | Includes bibliographical references and index.
Identifiers: LCCN 2023028663 (print) | LCCN 2023028664 (ebook) | ISBN 9780520380813 (cloth) | ISBN 9780520380820 (paperback) | ISBN 9780520380837 (ebook)
Subjects: LCSH: Communication in science. | Storytelling—Social aspects.
Classification: LCC Q225 .B57 2024 (print) | LCC Q225 (ebook) | DDC 501/.4—dc23/eng/20231012
LC record available at https://lccn.loc.gov/2023028663
LC ebook record available at https://lccn.loc.gov/2023028664

33 32 31 30 29 28 27 26 25 24
10 9 8 7 6 5 4 3 2 1

Contents

Illustrations and Table

FIGURES

TABLE

Acknowledgments

In September 2017, I gave a talk at UNLV's University Forum Lecture on "The Science of Stories and Stories about Science." At the time, the talk was no more than a collection of musings I had on the similarities between climate change and evolution. After the talk, my friend and colleague Jenny Guthrie asked, "So, you're turning this into a book, right?" I am so thankful for her encouragement to pursue those ideas further and that seven years of writing later, that book is finally here.

This book would not have been possible without the unwavering support of my husband and best friend, Paul La Plante. Not only did he listen to tedious hours of me brainstorming chapter ideas and talking through examples, but he also created a program (and taught me how to use it) to design the narrative webs. I can always count on his intelligence and practicality to walk me through any problem, academic or otherwise. Paul, I love you more than words can say.

I always want to thank my family. To my mom and dad, I am the woman and writer that I am today because of your support and guidance. You always let me be creative and allowed me to follow my dreams, however impractical or unlikely—thank you. To Graeme, Cheryl, Trevor, Molly, Charley, and Lisa—you are amazing siblings and friends who always lent a listening ear during our weekly family chats, started at the beginning of the pandemic, and celebrated every milestone of this book's journey with me.

The mental, emotional, and physical process of writing this book was supported by so many people (and a few furry companions). My cats, Ada and Izzy, have been my persistent writing buddies and lap warmers since we adopted them in 2020. I was also accompanied near-weekly by virtual and in-person writing buddies who helped me stay on track and bounce ideas off them while writing. My writing group companions over the past few years have included Leah Ceccarelli, Justin Eckstein, Payal Sharma, Eda Anlamlier, James Wynn, John Gallagher, and Aude Picard—thank you!

The team I worked with at UC Press has been absolutely outstanding. Thank you to Stacy Eisenstark who first saw promise in my book and shepherded me through the first draft. Thank you to Chloe Layman who fearlessly took on the project and was incredibly generous with her time, advice, and feedback. I also want to thank the members of the editorial staff at UC Press, including Chad Attenborough, art designer Lia Tjandra, Francisco Reinking, Linda Gorman, the anonymous reviewers, and the members of the Editorial Committee for the care and time they put into the book's content, design, and structure.

I also want to thank my extended academic family at UNLV and beyond. I am forever grateful for and eternally lucky to have colleagues and friends who have supported me as a scholar, writer, and person. This list of the many amazing people in my life is not exhaustive, but I do want to thank Arya Udry, Kaitlin Clinnin, Denise Tillery, Stephanie Willes, Robert Burgy, Alain Bengochea, Katie Walker, Jessica Teague, Tara McManus, Rebecca Rice, Natalie Pennington, Melissa Carrion, Chris Harris, LeeAnn Sangalang, Marcia Allison, Brooke Wolfe, and Randy Lake. A special thanks goes out to the members of the PhD Workshop I taught at Aarhus University in the spring of 2023 who gave me feedback on the potential application of narrative webs and helped me clarify key explanations.

Finally, I want to thank everyone who is reading this book and engaging with what I think are some of the most important issues of our time.

Whether you research or work in these areas directly, want to talk to friends and family about them, or are simply curious, I hope you take away something from the book and keep working toward making the world a better, more equitable, and sustainable place.

In memory of Pauline Harry, my family's best storyteller.

Introduction

REWRITING THE NARRATIVE OF SCIENCE COMMUNICATION

On April 23, 2020, then–US president Donald Trump addressed the nation about the COVID-19 pandemic. In response to comments from William Bryan, acting head of the US Department of Homeland Security's Science and Technology Directorate, who shared findings from a new study that said that heat and disinfectant kill the virus, Trump proposed that "a tremendous ultraviolet" light could be brought "inside the body" to cure people infected with the virus. He added that disinfectant should be explored "by injection inside" to act as a "cleaning" of the body and lungs.[1] These brief comments, which Trump quickly walked back as sarcastic, resulted in an uptick in calls to poison control centers and reports of people ingesting bleach.[2] Maryland issued a public alert about ingesting disinfectants after receiving more than one hundred calls in a single day, and New York's poison control center saw a doubling of disinfectant exposure immediately following the briefing.[3] A survey conducted in June 2020 reported that 8 percent of people surveyed said that they had consumed or gargled with bleach or disinfectants to prevent contracting the coronavirus.[4]

Words matter. The words of our leaders can influence people's behaviors, including their willingness to experiment with dangerous remedies.

When people have incomplete knowledge of a situation, we turn to those with more expertise, or perceived expertise, for guidance. Trump's COVID-19 briefings, held through the end of his presidency, updated the public on the coronavirus, but also spread misinformation about proper preventative measures and cures. When we hear about scientific topics in the news, online, or in conversations with friends and family, mixed messages abound. In the case of COVID-19, Trump's political authority frequently opposed the scientific and medical authorities of White House infectious disease expert Dr. Anthony Fauci and the Centers for Disease Control. In the case of climate change, scientific outliers are given equal talking time to climate advocates on news programs, infusing skepticism with near equal authority to the overwhelming scientific consensus. These communication conflicts reflect the various stories circulating about scientific topics that compete for attention and adoption with the power to guide public decision-making.

Overall, studies show that most people place either "a great deal" or "a fair amount" of trust in scientists and scientific institutions,[5] but this faith becomes muddied when specific topics such as climate change, which can be more polarizing, are discussed.[6] For some, science is an immutable source of truthful information, but for others, it is one of the most corrupt and powerful institutions of our time. For me, science is the evolving force of inquiry, experimentation, observation, and interpretation that constantly refines our understanding of the world. While science is most accurately a method or approach to problem-solving and information-gathering about the world around us, we might refer to *science* more generally as an institution in which scientific processes occur. One might say, for example, "Science demonstrates that climate change is real and dangerous," referring to a consensus of scientists or even scientific institutions as speaking for *science* as a method of inquiry.

The term *science* evokes one version of scientific thinking that privileges Western Enlightenment perspectives that emphasize empiricism, generalizability, and universality. However, Traditional Ecological Knowledges or Indigenous Knowledges are also science, in that they are also methods of inquiry rooted in lived experiences, ancestral knowledge, and observation. In this book, I use the term *science* to refer to institutionalized, Western understandings of science, or what has been called "Eurocentric science."[7]

In expanding our notion of science to include more than Eurocentric perspectives, we make space for both Eurocentric and Indigenous ways of knowing, which Albert Marshall (Mi'kmaq) called "two-eyed seeing."[8] To recognize multiple forms of scientific inquiry is to expand our understanding of where information comes from and the various ideological and historical influences that affect our conclusions.

Scientific information in all its forms is up against powerful forces of misinformation, conspiracy, oppression, exclusion, and minimization that not only affect how we react to specific controversies but also the trust we are willing to put into sciences as explanatory resources. The proliferation of science denial and skepticism has captured much scholarly and media attention due to its disastrous effects on public education, knowledge, and health. While some conspiracy beliefs may seem harmless, the spread of misinformation can pose formidable challenges to science communication and can even lead to death, as we have seen in the case of COVID-19 skepticism.[9] A common problem for science communicators is how to make complex, technical, or otherwise restricted information accessible and relevant for public audiences.

I consider the processes of sciences to be rhetorical and communicative practices.[10] By that, I mean that the ways we come to know information, gather evidence, interpret data, and share that information with others are rooted in human decision-making, language (and other symbols), and social influence. This is perhaps easiest to see in Indigenous sciences, because their methods are embedded in relationships, community knowledge, and personal experiences.[11] Instead of viewing nature as an object to be interpreted and analyzed, Indigenous knowledges view it as interconnections to be read, felt, and understood through embodied experiences.[12] In Eurocentric sciences, however, there is the systematic removal of the "human" in favor of the objective, unbiased, and removed observer.[13]

Science is rhetorical because scientific information and conclusions are shared purposefully within and outside of the institutionalized scientific community to make arguments about discoveries, interpretations of data, and new findings and gain adherence to those arguments. Communicating scientific information to public audiences involves strategic choices and audience adaptation. Rhetorician of science Leah Ceccarelli, for example, argued that Fauci had to manage the challenges of a Trump presidency by "selecting carefully from the available means of persuasion" to provide public audiences with accurate knowledge that would be "accessible" to nonexperts.[14] Films and television shows such as *An Inconvenient Truth* or *Years of Living Dangerously* translate technical information about climate change into real world examples.

Philip Wander, an early rhetorician of science, writes that scientists "persuade one another" through

> grant proposals, journal articles, and convention papers. . . . In order to be successful, they must convince [a professional] audience that the research topic is worthy of study, the appropriate tools were used, and used correctly, and that the researcher knew what he or she was doing.[15]

In addition to publications, many of the regular practices of science, such as applying for grants and writing articles, involve decision-making, negotiation, choice, and deliberation. A 2019 article published in *Nature* by Emily Sohn advertised the "secrets to writing a winning grant," which included advice on audience adaptation (e.g., "[do] an extensive search of available grants and not[e] differences in the types of project[s] financed by various funding bodies"), style and delivery (e.g., "use clear language instead of multisyllabic jargon"), and self-promotion (e.g., "a grant is about convincing a jury that your ideas are worthy and exciting"). Sohn also emphasizes the value of storytelling, noting that "personal details" and "conflict and resolution" can help reviewers connect to the value of the study and the people conducting it.[16]

Scientists choose what to study, how to design experiments, and what hypotheses and research questions to develop, all of which are embedded with the rhetorical dimensions of selecting certain reflections of reality over others.[17] Additionally, institutional pressures and perceptions may select certain scientists for some opportunities, lead to additional grant money or publications, or construct barriers to success in the job market and in the tenure process. Put in a different way, each choice within the practice of science is also a nonchoice, or an opportunity, a question, an experimental design not taken. Consequently, the process of science involves the attention and decision-making of individuals often operating within an academic or research institution, which may have its own set of practices, norms, and expectations that restrict the scientific process. If we consider science to be a rhetorical practice, influenced by choices and the communication of them, we open ourselves up to integrating rhetoric's tools and strategies for the benefit of science communication.

A TURN TO STORY

Of the many rhetorical strategies available for science communicators, I argue that science communication can be strengthened through viewing the process as storytelling.[18] Communicating science means expressing the collection of stories that constitute our ways of knowing the world around us. There are many definitions of what counts as a "story." To start with perhaps the simplest definition, English scholar Martin Kreiswirth defined a story as "someone telling someone else that something happened."[19] Things happening are not stories, but that humans recount events happening are. This "minimal" definition includes an event happening and then someone communicating its occurrence.[20] Rhetorician Chris Ingraham noted a similar definition of narrative, noting that "a segment of language" is a "'minimal' narrative" if it has "two temporally ordered narrative clauses."[21] This definition adds a temporal element between two events instead of Kreiswirth's single "something."[22]

In another broad definition, English scholar H. Porter Abbott focuses on grammatical structure to propose that "as soon as we follow a subject with a verb, there is a good chance we are engaged in narrative."[23] Abbott's definition emphasizes action occurring and that the action is being recounted through someone describing it through language. When I define *story*, I often turn to the definition offered by rhetorician Walter Fisher, because it captures the importance of sequence, action, and audience. Fisher defined story as "words and/or deeds—that have sequence and meaning for those who live, create, or interpret them."[24] Fisher's definition highlights the resonance of stories with audiences and how meaning is interpreted from a series of linked words or actions. In the spirit of broad definitions that can encompass a variety of storytelling practices, I define *story* as a form of symbolic or material communication that provides meaning for audiences and contains to greater or lesser degrees the components of character, action, sequence, scope, storyteller, and content. This definition captures the key features of other definitions while centering a story's natural variability in terms of structure, emphasis, and function.

While we may not immediately think of science communication, such as a science lecture in a classroom, as a story, we would likely consider an entertainment film a story. Consequently, some people may balk at the idea that science communication is a form of storytelling and may wish to separate the two as distinct practices. This perspective proposes that the falsities and exaggerations of stories have no place in "real" science. In a blog post about the relationship between science and stories, library consultant Thomas Basbøll argued that the power of stories was the primary reason he was "suspicious of storytelling as a means of conveying scientific ideas."[25] Other scholars believe that to leverage storytelling for the communication of science is to corrupt, distort, and compromise science's truth, objectivity, and purity.[26] Thus, they argue that narrative should be used cautiously, sparingly, or not at all in science communication.

When I conceptualize the "stories" of science, I am referring to how, whether giving a press conference on COVID-19 protocols or publishing an article about a new finding, scientific discourse always has narrative features such as a storyteller, content, a setting, an audience, sequence, and meaning. Although these features may be used in various ways and emphasized to different degrees, there is natural overlap in the communication of scientific information and other topics, such as the news, personal experiences, or other interactions. Indeed, Fisher described humans as *homo narrans*, or the storytelling animal, to emphasize the ubiquity of narrative forms in human communication.[27] Attending to how narrative elements are used—and abused—can improve how scientists share their insights and encourage the recognition of the storied elements within science communication.

I view narrative as a powerful vehicle for conveying scientific information. I encourage science communicators to employ narrative features in varying ways that are sensitive to the story's context and audience. Navigating these various tensions is difficult, and yet we strive for a way forward in the morass of scientific misinformation. I hope to challenge the assumptions of cynics who would separate science and story to retain science's purity. Instead, I advocate for the recognition of story as integral to the communication of science that we ignore to the detriment of both accuracy and effectiveness. Before elaborating on tools I have developed to help produce and critique science communication through a narrative perspective, I further explore the relationship between science and stories.

THE POWER OF STORIES

Narratives serve a variety of functions; they explain, explore, entertain, build community, and unite people against others who might tell a different story. Stories are a manifestation of our values and beliefs and are made up of ancestral knowledge and linkages to a community's past and reflections of human motivations, drives, understandings, and creativity. Conveying information through narratives not only improves understanding, comprehension, and retention of information,[28] but also leads to changes in beliefs, attitudes, and behaviors.[29] Stories put information into a format that people are familiar with and thereby increase retention and engagement. In a meta-analysis of studies on assessing narratives in science and health communication, Matthew Z. Dudley and colleagues found that narratives were effective across most of the studies in creating "openness to information by delaying the formulation of counterarguments" or resistance to the information.[30] Consequently, they propose that science and health communicators should use "story elements to improve engagement and influence" with public audiences.[31]

Research shows that stories are simpler, more engaging ways to transfer technical information on health and scientific topics and can increase identification with alternative groups, which refers to how much common ground you find with others.[32] Additionally, storytelling can avoid what are called boomerang effects, where audiences double down on previously held beliefs when faced with information that challenges those beliefs. Imagine that someone does not believe in climate change, so you give them information about how climate change is real and severe. A boomerang effect then occurs if the person, now given information contradictory to their original belief, is even more certain that climate change is not real.[33] This effect, in part, explains why simple increases in information do not always shift people's perspectives and can potentially backfire.

Alternatively, when we adapt the same scientific information to a narrative format, we are packaging the content in a way that helps audiences be receptive to our messages. For example, Angeline Sangalang and I studied the effects of different narrative structures on environmental attitudes and behavioral intentions, concluding that narratives can adjust people's willingness to engage in environmental action despite political affiliations that may initially deter environmental interest.[34] In a different context, Hyun Suk Kim and colleagues found that the presentation of information as narratives positively influenced intentions to quit smoking.[35] These studies, among many others, point to the effectiveness of narrative communication and its benefits regarding instruction, persuasion, and engagement, particularly for scientific, environmental, and health-related topics.

These findings are also borne out in our everyday lives. Early years of schooling infuse education into fun storylines through books and children's entertainment media such as *Sesame Street, Barney & Friends,* and *Peppa Pig.* Institutes such as the Norman Lear Center advocate for and analyze entertainment education for all age groups across a variety of media outlets. *Contagion,* for example, a 2011 Steven Soderbergh film about a global epidemic caused by a disease transferred from animals to humans, enlisted multiple scientific advisors and public health consultants to create an accurate and now quite prescient scenario.[36]

Stories represent broader social consciousness and provide insight into how people move through the world. Humans are not Spock-like characters who only make decisions through logic and pure rationality. Instead, humans are complex beings who also consider values, morality, and stories as decision-making tools that can guide us through everyday interactions, whom to vote for, and what life choices to make.[37] For example, folklore, cultural traditions, and histories inform beliefs and behaviors. Variations on the story of Santa Claus guide different Yuletide traditions, including if milk and cookies or mince pies are left out, whether presents are left under a tree or in children's shoes, and even on which days festivities occur. From the mostly private occurrence of family traditions to larger cultural patterns of trust in science, the stories we tell are pervasive and influential. In this sense, narrative is "an instrument of power"[38] in directing ways of thinking of and interacting with others and the world around us. To examine stories, therefore, is to examine the vehicle through which people and our relationships are constituted and reformed, and may potentially change.[39]

Stories are beautiful in their variety and pervasiveness. From the Brothers Grimm's fairy tales to autobiographies woven into Navajo blankets,[40] and from hula choreography in Hawaiian storytelling traditions (ha'i mo'olelo)[41] to movies from Hollywood, Bollywood, Nollywood, and other film industries,[42] stories are everywhere and come in many shapes and forms. Science communication, for example, recounts events, and contains characters, rising and falling action, scenes, and conclusions or morals. Science's stories inform how to make sense of the world and our place in it. Many other stories perform this function, such as folklore, religious parables, and political campaigns, and have their own methods for verification. The stories of science are supported by the verifiable and repeatable practices of scientific methods, which are validated through measures such as peer review, personal experience, and credibility. These tools, of course, are not foolproof; there are many examples of how these Eurocentric processes have led to medical injustices, exclusionary practices, and hoax publications.[43] However, scientific practices do provide a baseline of reliability that communicates, as best we know at a particular time, accurate information about humanity and the Earth.

The communication of science can sometimes lead to misunderstandings, confusion, and controversy, which imperils building an open, accessible, and inclusive scientific community. Such distortions have occurred, for example, in the human origins controversy, where creationists purposefully misconstrue scientific disputes over data as evidence of science's inability to validate evolution.[44] Similarly, Andrew Wakefield intentionally leveraged vague language such as "may be" and "possible" and the use of the passive voice to suggest a relationship between vaccines and autism diagnoses without demonstrating a relationship through evidence.[45] These stories were told with the intention to distort and misrepresent scientific information. Although some would wield storytelling for nefarious purposes, we would be mistaken to attribute these unethical qualities to the narrative form itself and not to those using it. If narratives are tools malleable to the intentions of the storyteller, then one must question why scientists and science communicators would eschew such a valuable tool because of how others have used it. It is perhaps unethical, then, to reject storytelling as a tool for science communication, because it unnecessarily provides an unchecked tool to those who would wield it to damage public well-being.[46]

For portions of the population, scientific stories are reasonable, credible, and meaningful. However, narratives that "give solace to some . . . will remain forever unsatisfying to others."[47] While science's stories are important sense-making tools to understand reality, different stories and storytellers may produce different levels of adherence in audiences. In April 2020, for example, some people interpreted Trump's simple problem-solution story of COVID-19 being cured by bleach as authoritative, albeit from a political authority rather than a formally scientific one. For others, the ludicrous statement was not science at all—it lacked evidence, testing, experimentation, and a fundamental understanding of the toxicity of bleach. Therefore, it behooves us to take a closer look at the type of stories told about science, who might be excluded from these stories, and how we might tell better ones.

THE STORIES SCIENCE TELLS

When we think of "science" being conducted, we may instinctively conjure up stereotypical images of White men in lab coats using pipettes or microscopes. Science, however, can be conducted in a variety of different places by a variety of people. Attempts to restrict science to certain actions and exclude others are gatekeeping practices that sociologist Thomas Gieryn refers to as "demarcation."[48] Demarcation is a rhetorical practice that negotiates the boundaries of science, deciding what gets to count and what does not. Demarcation can be a helpful practice, such as the cultural distinction between medical treatments and pseudo-science. For example, we may demarcate COVID-19 vaccines and Paxlovid as scientifically verified preventative and curative treatments for COVID-19 while we demarcate untested and unproven "miracle cures" such as ivermectin and chlorine dioxide outside of verified science.[49]

Practices of demarcation can also be harmful and exclusionary. Astrophysicist Chanda Prescod-Weinstein argues that science is dominated by "white empiricism," which validates only certain types of knowledge as rational and gatekeeps certain bodies—namely nonwhite, nonmale bodies—from its endeavors.[50] Technical communication scholar Natasha N. Jones notes that scientific gatekeeping upholds a "false dichotomy between who can do and communicate about science" that fails to acknowledge the performance of scientific processes within marginalized and nonexpert communities.[51]

Similar issues of expertise emerged in the controversy over the Flint Water Crisis. When Michigan switched its water source to the Flint River in April 2014, there were immediate consequences for residents who were exposed to toxins in the water. Locals were at a power disadvantage by not being formal experts in water safety, which was further compounded by the primary figures in the movement being women and mothers, of whom many were people of color living in low-income communities.[52] Flint activist LeeAnne Walters, for example, entered scientific spaces by documenting her home's water quality, researching the health effects of lead poisoning, and demanding water quality reports be conducted to understand the negative health effects experienced by her child. Walters was one of the many women in Flint who leveraged technical information with their personal accounts as mothers to lobby Michigan officials. Despite not being formally recognized as "experts" or authorities, women activists, such as members of Water You Fighting For?, mobilized for environmental justice concerns that affected their families' health.

Women are often excluded from gaining access to formal institutions and credentials and consequently must go to extreme measures to be recognized as scientists. Many scientists, such as Rachel Carson, sought employment outside of academia due to gender-based exclusion and discrimination.[53] There are myriad stories of women scientists who were denied credit on discoveries and Nobel Prizes, such as Cecilia Payne, Chien-Shiung Wu, Vera Rubin, Jocelyn Bell-Burnell, and Lise Meitner, due to committees overlooking and undervaluing their contributions.[54] If these are the stories institutionalized science tells, then we should want to write better, more inclusive stories that recognize the breadth of doing science and being scientists.

Demarcation practices are a continuous issue for Indigenous communities, whose Traditional Environmental Knowledges challenge Eurocentric science's exclusive access to knowledge. Institutional climate researchers have begun taking Indigenous science and methodologies into consideration in tracking changing environmental conditions, animal migration patterns, and ecosystem relationships to provide a fuller picture of our changing climate.[55] Bridging gaps between mainstream scientific communities and other scientific communities challenges previous demarcations that can often lead to mistrust or the colonization of ideas and knowledges, but is also valuable, integral work in developing and diversifying scientific processes.[56]

Demarcations in science are not fixed but are ongoing negotiations. What counts as science, and what is considered "normal science," is what is "espoused by a given scientific community at a given time."[57] Thomas Kuhn, a philosopher of science, argued that scientific disciplines are "characterized by continual competition between a number of distinct views of nature" that are negotiated by the scientific community.[58] Scientific information does not simply accumulate over time but is the product of controversies, disagreements, "accidents," and "anomalies" that lead to "scientific revolutions."[59] Scientific revolutions are paradigm shifts that upend previous ways of knowing and may redraw the lines of what is considered verified, "normal science," and what is outdated belief.

For example, in the Michelson-Morley experiment in physics, A. A. Michelson and Edward Morley sought to measure "luminiferous ether," the undetected substance through which scientists theorized light traveled. The experiment produced what has been called "the most famous null result in the history of physics,"[60] providing no evidence for the existence of ether. Despite this result, Morley was not convinced and continued experimenting, hoping to find a way to measure the ether, to no avail. The null result eventually influenced the development of Albert Einstein's special relativity to explain the movement of light. Michelson and Morley did not incrementally add to our knowledge of physics but instead contributed to the overhauling of our understanding of light to pave the way for new, more accurate knowledge to develop. Through a scientific revolution, "ether" became relegated to pseudo- or even nonscientific belief, redrawing the lines of demarcation. Demarcation is not inherently harmful and is indeed integral to the process of science, but it can also be wielded as a gatekeeping tool to preserve the perceived purity or objectivity of science and to discipline beliefs deemed outside the realm of science.

This book charts the various narrative features of texts from mainstream scientific sources, institutions, news media, and other discourses circulated in print and online around scientific topics, including from skeptical perspectives and perspectives otherwise demarcated outside the realm of Eurocentric science. Locating patterns helps us develop a deeper appreciation for the storytelling practices of science communicators, who operate in a variety of spaces, disciplines, and perspectives. I propose that strengthening the stories that science communicators tell and marshalling a broader definition of who counts as a science communicator are promising and effective ways to engage people in the conclusions and methods of scientific inquiry. Instead of strict binaries and exclusive categories, we should turn to more inclusive, holistic, and adaptable methods of storytelling. The process of incorporating narrative features can be facilitated through the visual mapping tools I call narrative webs and narrative constellations.

MAPPING CONSTELLATIONS ON NARRATIVE WEBS

Communicating complex scientific topics can be difficult, so scholarly work in science communication often provides practical advice and strategies. Many studies of best practices in science communication, including some of my own past work, have focused on simplicity, directness, and practicality for "general" audiences or that work across large swaths of people. Unfortunately, a focus on generalizability can have the inadvertent effect of universalizing audiences into a neutral, abstract community that is being addressed, which is not representative of the public's heterogeneity. In other words, single-message designs and best practices are often aimed at what will work for the "majority" of the public, namely dominant groups such as White, heterosexual, middle class, able-bodied men. Therefore, to develop best practices, there is an inherent, but not typically intentional, exclusion and marginalization of voices and audiences outside of the majority or perceived "neutral" audience. In my offering insights for science communication in this book, it is important to note that there is no "one size fits all" story that science communicators should adopt. Instead, rhetorical theory advises that communication should be tailored and specific, given circumstances, audiences, and contexts.

Science communication thus faces an important tension between universalism and personalism. If we universalize an audience, we reduce differences, eliminate power dynamics, and minimize context to a perceived neutral space. Alternatively, if we focus on differences too much, we may never locate any ways forward that can be generally adhered to or applied in the practice of science communication. In splitting this difference, my first book proposed a meso-level engagement whereby individual conversations can inform patterns to help tailor messages at the group level.[61] Entrenched in a meso-level approach are the opportunities and risks of both universal and personal level engagement, necessitating a balance and keen attention to influencing factors such as audience, purpose, and circumstances.

Science communicators face the monumental task of presenting information in a way that makes sense to nonexperts given a variety of different communicators, topics, factors, circumstances, audiences, and obstacles, while negotiating the tensions of universalism and personalism. To help strengthen science communication by attending to its narrative features, I propose the analytical tools of "narrative webs" and "narrative constellations." Narrative webs are a visual mapping tool that breaks down stories into six narrative features (character, action, sequence, scope, storyteller, and content) and maps them along three rings from the narrow and specific (micro-ring) to middling specificity (meso-ring) to the broader and more abstract (macro-ring; see figure 1). For example, a micro-ring character in a narrative may be a particular person, while a story with a macro-ring character may have no people at all and may instead focus on abstract processes and chemical interactions. When mapped on the web structure, the shape formed is what I call a narrative "constellation."

Narrative webs are an alternative to a linear scale or rigid binary between story and nonstory that recognizes stories as complex and flexible. A binary scale would simply group stories into two groups: a fairy tale is a story, and a recipe is not a story. A linear scale would account for gradations of story from a nonstory to a story based on the addition of specific elements. Consider a recipe blog that includes a recipe for how to bake a cake but also includes reflections from the chef about the baking process, experiments they did with the ingredients, and the special occasion where the cake would be served. This blog may be mapped more story-like than

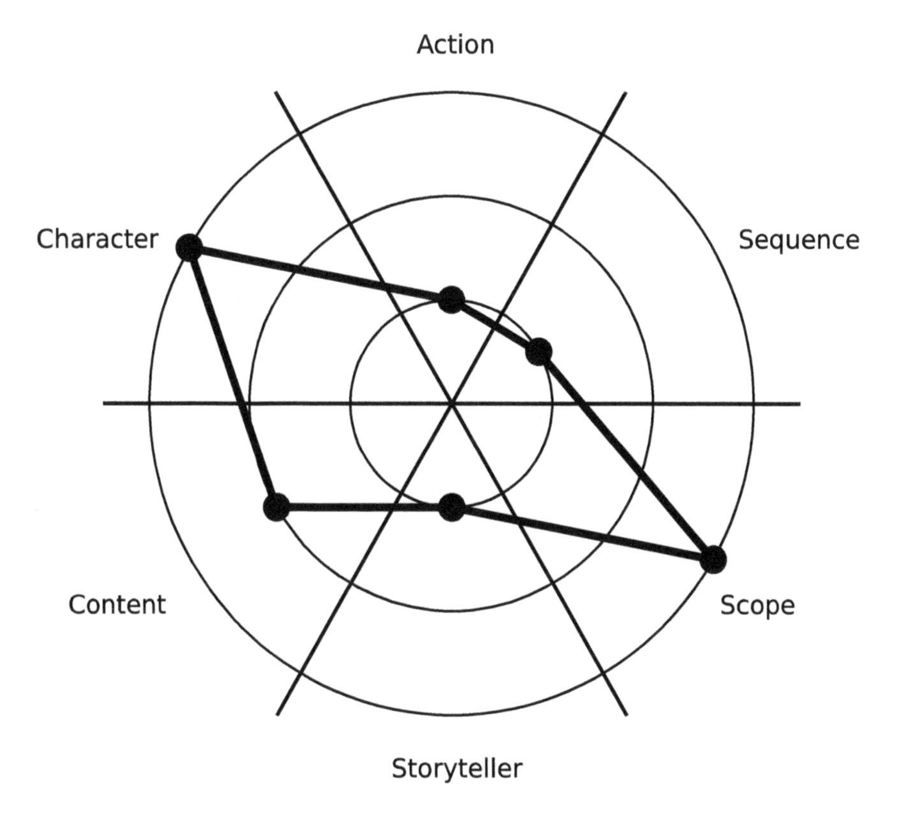

Figure 1. An example of the narrative web structure with a constellation mapped by connecting the dots plotted on each narrative wedge.

a plain, reflection-less recipe, but perhaps still not as story-like as a fairy tale. Alternatively, narrative webs evaluate narrativity across six different features, indicating that some aspects of storytelling will more closely resemble expected narrative elements while others will be deemphasized or appear less story-like. The three examples of a recipe, recipe blog, and fairy tale could be evaluated as to the extent to which they contain detailed, specific, and relevant characters, actions, sequences, scopes, storytellers, and content. Narrative webs account for additional gradation and flexibility beyond reductive science versus story binaries and instead attend to how we might evaluate a story's components.

Every story will have unique components, meaning there is no universal narrative form. Even so, I argue that narrative webs are a useful, visual

Table 1 Brief Descriptions of the Narrative Wedges

Narrative wedge	*Description*
Character	Agents, subjects, and objects involved in the action of the story
Action	The plot of the story, what characters are doing, the events of the story, and the likelihood of events unfolding as described
Sequence	Ordering of the story's events; its chronology, progression, and arrangement
Scope	What falls under the purview of the story in terms of space and time
Storyteller	Who or what is narrating the story, distributing the story, or recounting the story
Content	The topics of information discussed in the story and whether they are relevant to and resonate with the lives of audience members

heuristic to evaluate a story's narrativity for individual assessment, comparison, and identifying opportunities for modifying its components. At the very least, charting stories using narrative webs can help ensure that one is using a variety of storytelling features in one's science communication that will be relevant and interesting to target audiences. Using narrative webs acknowledges that including, emphasizing, excluding, and downplaying narrative elements are all rhetorical choices made in presenting a story.[62]

In a narrative web, each segment between the spokes represents a different narrative element that shows how the feature is being deployed, from the concrete to the abstract, which I refer to as a "wedge." The six wedges are character, action, sequence, scope, storyteller, and content (see table 1). The character wedge refers to the agents, subjects, and even objects involved in the story. The action wedge refers to the plot, what the characters are doing, and the events that occur in the story. The sequence wedge refers to the ordering of events and the relationship between the past, present, and future in the story. The scope wedge refers to what is included in the story or falls under its purview, primarily in terms of geographic distance and time scales. The storyteller wedge refers to who or what is narrating or sharing the story or is named as the source of the story. The content wedge is related to the topics or information covered in the story and its relevance to and resonance with audiences. In the following chapters, I discuss the wedges in pairs that are most closely related to one another (character and action, sequence and scope, storyteller and content), but all of the wedges are interconnected and influence one another to create a coherent whole.

I call the structure a narrative "web" because of its similar appearance to a spider web. Instead of separate lines that mark the narrativity of each feature, I wanted to combine them to emphasize their interrelationship and to capture visually the rings' relative narrowness or breadth. This shape also conjures up Abbott's description of narrative definitions as "quickly becom[ing] a tangled web."[63] The structure of the spider's web contains strands that radiate from a central space called the "hub spiral." The hub spiral is where the spider builds from to make the rest of the web and where it waits for prey to get caught. In a narrative web, the hub spiral is the micro-ring of specific, relevant features, which acts similarly as a foundation through which the rest of the constellation can be built. A story that lacks any feature on the hub spiral may be difficult to comprehend because it would lack specific characters, actions, sequences, scopes, storytellers, and content for audiences to relate to. Alternatively, locating story features solely around the hub would not create very captivating stories as there would be no strands extending the web's length to provide variety. Instead, I recommend mapping narrative features to create a constellation shape of interconnected features with at least two anchored on the hub spiral to create a varied story to "capture" the attention of audiences (or flies).

There are two primary uses of narrative webs: evaluating stories and creating stories. To use a narrative web to critique a story, one would categorize information one is receiving and plot the various elements on the wedges, attending to what is present, absent, specific, and abstract within the story. As an example of the mapping process, imagine there is a family who is having a conversation about how hot it has been, even well into the fall months. The father of the family, CJ, talks about his personal experience running a painting business and how his workers "cannot work full days because of the heat risk." He's having to take on fewer jobs, which is cutting into his profits and making it harder to keep the business afloat and thereby run the household.

CJ's daughter, Aanika, is a graduate student in environmental sciences at her local university and discusses the long history of changing weather patterns and how the longer summers are due primarily to climate change. Aanika widens the conversation to include not only the effects of the weather on their local community and her father's business but also how people and more-than-human life around the world are being affected by changing weather patterns. Luca, Aanika's brother, chimes in and says that climate change is "liberal propaganda," and climate scientists have been repeatedly wrong in predicting the end of the world. Instead, he listens to media and political figures whom he trusts more than scientists. He accuses Aanika of being indoctrinated and distracting from their family's plight.

Mapping the three stories of CJ, Aanika, and Luca about increased heat can be done on the same narrative web to compare their similarities and differences (see figure 2). While CJ is talking about a personal example (himself and his business), Aanika is addressing the more abstract characters of human and nonhuman life, and Luca is discussing a broader political group of people. CJ's scope is the smallest in examining a single business and the local community, while Luca is talking about US politics, and Aanika, the world. In terms of sequence, they are all talking about sequential moments in time, but Aanika is oriented toward imagining a more sustainable, but less tangible, future. As a family at the dinner table, all of the storytellers are well known and reasonably trusted. They are all discussing content relevant to the family, although CJ's discussion of money is arguably more relevant and pressing than broader political or environmental concerns to audiences who are, in part, relying on CJ's income to support the family.

Based on how broad Aanika's constellation is compared to CJ's mapped tight around the micro-ring, the narrative web in figure 2 illustrates how a specific and relevant story like CJ's can be captivating for audiences. Alternatively, Aanika's story has only one feature mapped on the hub spiral, meaning that the story does not have a firm anchor for audiences to relate to beyond their appreciation of Aanika as a storyteller. Luca's and CJ's stories also map the storyteller on the hub spiral, but they additionally make use of other narrative elements to add specificity and relevance to the story. I am not advocating that we only tell stories like CJ's or Luca's

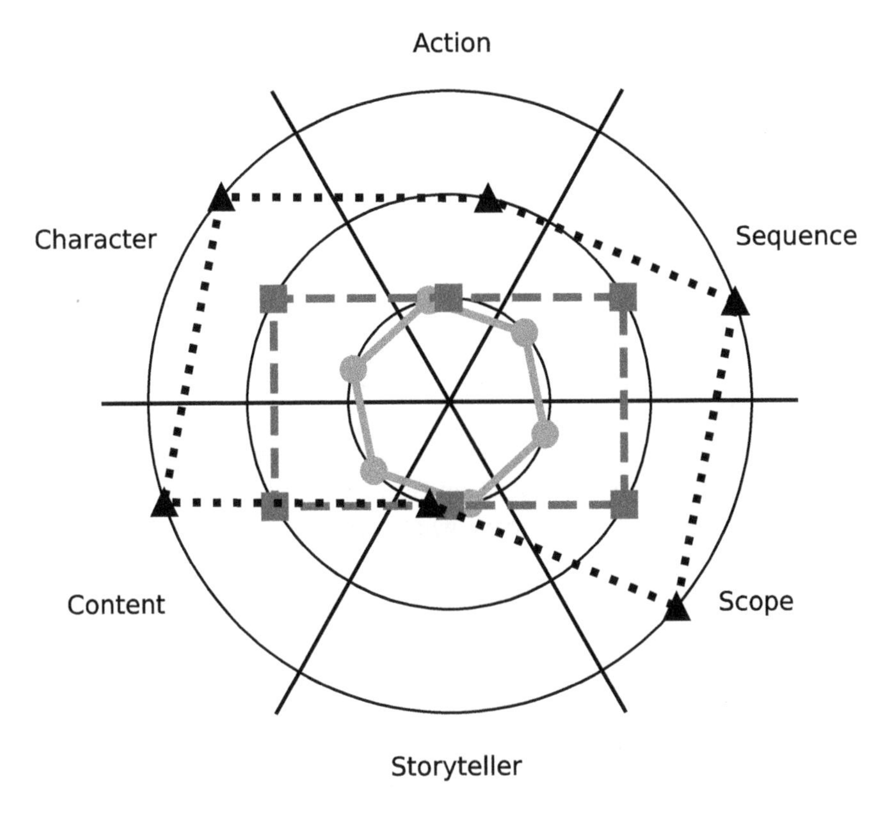

Figure 2. A narrative web of the family discussion about increased heat in their community. From most to least concrete, CJ's story is mapped using a solid line, Luca's story is mapped using a dashed line, and Aanika's story is mapped using a dotted line.

but that we may borrow from the specificity of the micro-ring to create anchors for more creative and dynamic storytelling that supports scientific knowledge and advocacy.

Plotting more wedges on the micro-ring or at least being aware of a story's narrative shape may help storytellers who may tend to focus on abstract scientific findings. Aanika, for example, could have included a personal experience with the environment to add a narrowed scope or discussed how increased heat will only continue to cost people like their family money, increasing the relevance of the story's content. Aanika could have also connected global environmental concerns to the family's backyard or brought in a specific character to emphasize who is involved in the story, such as research she is doing or content she is learning in her classes. Similarly, CJ's story could have included a broader sequence to discuss how the family might be affected in the future beyond the present moment or expanded the characters to address outdoor workers as a whole who are all affected by increased heat. Mapping these stories together creates a visual reference for how their narratives differ between various features. Each family member is telling a certain story they believe will be effective and that represents their perspectives, values, and priorities, but they are emphasizing narrative features in ways that may affect audience adherence.

Science communicators can also use narrative webs to develop scientific stories. Narrative webs can help science communicators reflect on how they are telling science's story, what aspects of the story they might change given its shape on the web, how the story compares to other circulating stories, and how the pieces of the story fit together along the web's wedges. To use a narrative web to construct a story, one would first need to identify their audience, communicative goal, situation, and any constraints that might influence the story's telling.[64] Then, one could outline the various components of information they want to share and see how it maps onto the various elements based on the elements' relative concreteness or abstractness. Afterward, they could locate potential adjustments to meet their circumstances, adapt to their audiences, and account for aspects of the situation. More insights about how to build scientific narratives with hub features, including how to create more dynamic constellations, will be explained with examples in subsequent chapters.

As with other analytical tools, I want to assure readers that there is no "right" way to map the above stories, or any story, on the narrative web. We do not want to use the narrative webs with the goal of being "accurate" in how the map reflects the story. Instead, narrative webs should be used as a tool to evaluate stories we or others are telling and how we might change them to create resonance with audiences, meet audiences where they are at, and compare the story to other stories. The tool relies on the interpretative work of the critic or story-builder to translate narrative features onto the web structure, which is itself a process of interpretation at the heart of rhetorical analysis.

I would happily engage in deliberation over how relevant or specific the content wedge is for each story in this example, which is something that changes based on audience and context. If the family discussion of rising heat were to have happened in a home in Las Vegas, for example, it would likely be highly relevant as seemingly endless stories are published on the risks of increased heat to residents and the shrinking water levels of Lake Mead. For others at a dining room table not immediately impacted by slightly warmer weather, however, threats to the environment may not seem as pressing or relevant as a family member about to lose their business. I would also welcome what would likely be fruitful discussions about the storyteller wedge and the relative trust the family might place in each other based on the stories they tell themselves. For example, Luca's consumption of particular media outlets may influence his views of Aanika as a rational storyteller, and Aanika may feel the same way about Luca.

In mapping and interpreting stories using narrative webs, it is important to take into consideration variations in effectiveness based on circumstances, audiences, and purposes. Narratives have three main purposes: "to entertain, to instruct, and to move."[65] It would be unreasonable to evaluate an entertainment film, which has the goal of entertainment, the same way we would analyze a classroom lecture, which has the goal of instruction.[66] For example, a chronologically ordered story, which would be a micro-ring on the sequence wedge, may be most suited to the purpose of explanation. But the same sequence of chronological order may be counter-productive when the underlying function of the story is to entertain or invert expectations, such as in a film or play.

One way to evaluate audiences' differing responses to narrative features is through Fisher's concept of narrative rationality. Narrative rationality is based on how "some stories are better than others, more coherent, more 'true' to the way people and the world are."[67] When faced with competing stories, people judge them and then decide to adhere to them or not based on the narrative criteria of probability and fidelity. A story's probability describes its internal coherence, or how likely the events in the story are to follow from one another. A story's fidelity describes its external resonance, or how faithful the story is to a person's conception of reality outside of the story. When a story has high probability and fidelity, the story makes sense to its audience, is persuasive, and will likely be adopted by them. Stories with low probability and fidelity may not be well understood and may risk being rejected as incoherent, irrational, or implausible in favor of other explanations. For the narrative web, narrative probability is primarily evaluated in the action wedge and narrative fidelity is primarily evaluated in the content wedge.

The storyteller can also affect a story's narrative rationality. For example, in the family conversation, someone hearing CJ's story may believe he is too close to the topic to be trusted and may be only looking out for his family's livelihood. However, others may view CJ as the most authoritative and knowledgeable due to his having specific, relevant experience regarding the topic at hand. Specificity and proximity can be interpreted differently by audiences to confer or remove authority and trust. For this reason, among others, it is important to diversify science communicators so that there are many stories and storytellers for people who may reject or who may be skeptical of official scientific voices.

In breaking down barriers in science and in stories, we can all be empowered to be scientific storytellers in our everyday conversations and, for some of us, in our more formal careers. Narrative webs can be used to see if storytellers are privileging certain perspectives at the expense of excluding more meso-ring and macro-ring storytelling elements or solely telling stories at a single register, such as in CJ's story. Alternatively, narrative choices that rely on universal ideals may limit the diversity, variety, and accessibility of science's story and reify a globalized narrative that some argue restricts public interpretation and could even be considered violent acts of cultural erasure.[68]

Therefore, the key to an effective narrative is a variety of different rings mapped with the consideration of target audiences and the storyteller's goals in relaying the information. I propose that when mapping a narrative science communicators should aim for a constellation pattern that places narrative features across different rings, as relevant to a target audience. At least two wedges should be mapped on the hub spiral, the micro-ring, to provide an anchor point for scientific stories that is specific, relevant, and easily understandable. The placement of dots across the narrative wedges as such creates the shape of a constellation. Narrative constellations celebrate diversity and variety in scientific stories as opposed to prescribing a universal or ideal story type.

The constellation is a useful visual heuristic and also a meaningful metaphor for establishing relationships and variety within stories. In an article about the #MeToo movement, I conceptualized a both/and approach to personalism and universalism through the metaphor of a constellation. Individual stars represent particular instances of sexual violence while the constellation as a whole gives a fuller view of the magnitude of the problem and the systems in place that enable sexual violence. Tarana Burke, the originator of the #MeToo movement, employs a constellation approach in noting that structural changes will be needed in order to address the pervasiveness of sexual violence, but also that attending to individual stories and experiences, and how sexual violence affects different communities at different rates, is also integral to the stories of #MeToo.[69] Instead of privileging or advocating for the individual or the collective solely, a constellation approach considers the value of multiple rings, especially together.

A narrative constellation does not aim to be universal, but is sensitive to individual differences, diversity, and the power of the individual. A rich variety of specific and more abstract features create a dynamic story that has specifics for audiences to grasp onto while still describing the full picture and playing with abstract concepts. Scientific stories that are constellations of narrative features will likely reach a wide audience because they balance the rich diversity and variety of storytelling techniques, traditions, and contexts to tailor stories to target audiences.

I use the concept of constellations to guide narrative theorizing because it emphasizes interconnectedness, relationality, and differing perspectives. In Indigenous and cultural rhetorics scholarship, kinship, reciprocity, and respect are integral values that reject hierarchies in favor of webs of relationships. The term *constellations* is thus representative of those interconnections and the valuing of multiple perspectives. Cultural rhetoricians view constellations as reflective of "build[ing] community and understanding" where research is done "in constellation with others," including other scholars and communities, to be "supportive of equity and change."[70] Constellations represent a critical, justice-oriented approach to storytelling that centers audiences, differences, and diversity among the narrative features.

For Leanne Betasamosake Simpson, a Mississauga Nishnaabeg scholar, constellations are "doorways" into other worlds that "reveal theory, story, and knowledge" through time.[71] Constellations are relational, "otherwise they are just individual stars," and thus represent networks of people brought together by storytelling, common histories, and collective struggle.[72] Vox Jo Hsu, drawing on Simpson's work, argues that narratives establish "relations and responsibilities" and proposes that stories can be understood as constellations that are in constellation with other stories.[73] Like Simpson and Hsu, I see much resonance between constellations and storytelling in terms of how storytelling features relate to one another and how we collectively relate to stories. When we tell stories, we are not simply shouting into the void; we tell them purposefully with the intention of reaching, influencing, and/or connecting with an audience. Science's stories are the same.

By viewing science communication as stories, we can understand why scientific messages have not been as persuasive for some audiences as science skeptical messages have been. Eurocentric science communication tends to minimize human influence, use passive voice, and contain expansive scopes, and may not have a clear ending or purpose. Thus, there are few to no concrete or "micro-ring" aspects of the story that resonate with audiences on a personal level. Inherent in each narrative feature are narrative opportunities, or places where science's stories can be deepened, modified, and extended to make our acts of science communication richer and more attentive to the needs and expectations of diverse audiences. To illustrate the value of narrative webs and constellations, I apply them to scientific controversies as nexuses of competing narratives from scientific and alternative sources. Many of these alternative sources are skeptical ones that may undermine scientific information or create roadblocks to meaningful action. Scientific controversies are not only some of the most pressing issues of our time, but they are also valuable case studies for exploring how to strengthen scientific communication in the face of competing mis/disinformation.

SCIENTIFIC CONTROVERSIES

This book explores four controversies throughout its chapters: climate change, evolution, vaccination, and COVID-19. I use climate change as the orienting case study to serve as a consistent starting point for exploring each of the narrative wedges. Additionally, climate change has been acknowledged as a "threat multiplier," meaning that climate change's effects exacerbate other problems and create instability across environmental, health, political, and social contexts.[74] A full consideration of each scientific controversy could easily be their own texts,[75] but I selected four example controversies so that we might start to trace patterns in how scientific stories tend to be told. The four controversies act as preliminary comparisons to analyze narrative features across scientific stories with attention to their deployment and dynamics within each controversy. As this book's analyses show, there are many similarities but also notable differences between the stories' narrative elements, which provide insight into the strategies available to science communicators and opportunities for diversifying standard practices in scientific discourse.

I analyze the chosen scientific controversies by collecting a variety of writings and explanations from science storytellers and locating commonalities and distinctions. For example, I examine public articles, news media, social media, books and textbooks, publications, messages, and content published by scientific institutions and political groups (e.g., content about climate change published on the Environmental Protection Agency's website and pamphlets from the Centers for Disease Control about vaccines). These are rhetorical "fragments" of the larger scientific discourse through which scientific topics are constructed and discussed.[76] These texts aim to capture the general substance and themes surrounding these topics. These fragments are just some examples of how these controversies manifest, but they provide useful inroads for exploring the dynamics of each controversy, how narrative features are deployed within them, and potential considerations for the future of science communication.

It is not enough, however, to only think of science communication in a vacuum; we must also acknowledge the relationships, interconnections, and tensions surrounding scientific topics in public discourse. Consequently, in addition to gathering narratives from institutions and official scientific sources, I was also interested in the broader conversations happening about these scientific controversies from alternative, public, and digital sources. These stories, which Fisher calls "rival stories," compete with scientific stories for adherence and attention and are part of the larger context that scientific stories circulate within.[77] Similar to collecting science's stories, I gathered fragments from prominent voices in public, news, and digital spaces to represent rival stories within each controversy. These gathered texts are not meant to be exhaustive or representative, but to provide a range of rival stories to locate patterns and differences. Each of these controversies affect human and nonhuman life and the very nature of knowledge, expertise, and decision-making to various degrees, making them integral components of public life and science communication. The next sections provide brief overviews of each of the four controversies, but more details on the controversies, including their rival stories, are provided in the next chapter.

CLIMATE CHANGE. There is perhaps no issue more pressing and existentially threatening than the book's primary case study, our changing climate. Already affecting the most vulnerable among us and with little progress in terms of global adaptation and mitigation, climate change's consequences are predicted to get even worse, potentially disrupting global stability and altering life as we know it.[78] At risk in this controversy is the irreparable damage of humans on the environment, climate wars, famine, drought, and the increase of extreme weather events, which all point to devastating changes in societies across the world, especially for marginalized communities.[79] Part of the climate change controversy stems from public misperceptions around scientific consensus on anthropogenic climate change.[80] Other issues emerge in the weighing of climate change's potential environmental consequences against making economic sacrifices that might damage businesses or deprioritize human comfort. A variety of factors have been attributed to causing policy delays, such as biased media coverage, the influence of industry and lobbyists on legislation, political and religious beliefs, and online echo chambers, among others.[81]

EVOLUTION. The controversy over evolution largely resides in whether evolutionary explanations can account fully for the level of intricacy found in contemporary life forms and whether scientific explanations preclude or exclude the possibility of a role for the divine, especially for Christian audiences. In this controversy, there are large educational stakes. As the foundation of disciplines such as biology, a belief in evolution can undergird understanding of larger scientific principles.[82] Links between doubting evolution and subsequently questioning climate change and vaccine safety show that evolution is a gatekeeping belief.[83] The National Science Teachers Association put it simply: "If evolution is not taught, students will not achieve the level of scientific literacy needed to be well-informed citizens."[84] Studying human origins remains of vital importance for science communication scholars because "questions about evolution have served as proxies in larger debates about religious, ethical and social norms" that drive society's operation and well-being.[85]

VACCINATION. The anti-vaccination movement started in the 1800s and has appeared cyclically in public consciousness. The movement reemerged in the 1990s from relative dormancy with the controversy over an article published in *The Lancet* that falsely claimed a link between the MMR vaccine and autism.[86] With declining childhood vaccination rates into the 2000s and 2010s, there are increased risks to children and for outbreaks, such as the measles outbreak in Disneyland in 2014.[87] The growth of the anti-vaccine movement also threatens the durability of herd immunity, which protects those who cannot be vaccinated from contracting contagious diseases. The movement has contributed to the global resurgence of previously eradicated diseases.[88] While it is difficult to quantify precisely the cause-and-effect relationships between the anti-vaccine movement and disease outbreaks, many have attributed the rise in anti-vaccine rhetoric with thousands of disease outbreaks and hundreds of deaths.[89] Consequently, the World Health Organization lists vaccine hesitancy as one of the most dangerous health risks of our time.[90]

COVID-19. Vaccine hesitancy also threatens the world's ability to respond to emerging pandemics such as COVID-19, which, at the time of this writing, is still causing thousands of deaths weekly. As of April 2023, there have been over one million deaths and nearly 103 million infections in the United States alone and nearly seven million deaths and more than 760 million infections worldwide.[91] The COVID-19 controversy involves vaccine skepticism, but is also animated by concerns of government overreach, mandated lockdowns, mask-wearing, and travel restrictions. The

COVID-19 controversy encompasses issues of misinformation, individual risk calculation, government mandates, and the ability of scientists and medical experts to respond to the unfolding crisis in real time. With changing advice and proposed measures, along with political polarization around the urgency and severity of COVID-19, there has been much outcry against preventative and curative measures as society tries to navigate the likely long-lasting effects of COVID-19 on public health and the politicalization of medical information and experts.

Scientific narratives and their rival stories illustrate the complex dynamics involved in science communication, evaluating credibility, and knowledge building. Within these dynamics, I am interested in locating ways to strengthen the rhetorical toolbox of science communicators. What kinds of stories are typically told about scientific topics? What differentiates scientific stories from rival stories? How can we leverage storytelling features to engage multiple, diverse audiences in science communication? How can we productively explore the various features of story through narrative webs? In this book, I begin answering these questions by exploring the nexus of science and story in contemporary scientific controversies.

CHAPTER OUTLINE

Chapter 1 provides more in-depth discussion of the four scientific controversies, their histories, how their stories have unfolded, and similarities and differences that create useful patterns in determining how to approach scientific communication. Chapter 1 also provides more details on the "rival stories" that science communicators are in conversation with from political, economic, religious, and conspiracy storytellers, as well as productive rival storytellers who contribute to diversifying science and breaking down harmful instances of demarcation.

Chapter 2 examines the tensions around characters and the actions they take within scientific narratives and rival narratives. I demonstrate how scientific stories typically downplay human characters and instead substitute them with "science" writ large, scientific disciplines, personifications of nature, and collective groups. These substitutions of individual agents can make scientific stories less appealing to audiences expecting individuals to be a story's main characters. However, focusing on individuals exclusively is not a cure-all for scientific stories. The stories of science are rarely stories of individuals, but of entire labs, groups, and communities who perform many types of actions that can be undermined or hidden by focusing on individuals, who tend to be White, male, and members of dominant groups. Science communicators and storytellers can use this chapter to adjust the character and action wedges to express scientific information in more engaging and personal ways.

Chapter 3 examines the wedges of sequence and scope. Spatial and temporal scopes that are too small or too large may be hard for audiences to comprehend, such as microscopic scales of individual genes to macroscopic time scales of billions of years. Through citing previous literature and providing examples, I demonstrate how scientific stories often contain scopes that are difficult to comprehend and employ narrative sequences that rely on the uncertainties of the future. These narrative decisions can make scientific stories less probable and coherent to audiences. Alternatively, focusing scopes only in terms of a human lifespan and frames of reference can do a disservice to science in appreciating its vast wonders and expanses and how tiny changes can have large impacts. While telling stories with clear beginnings, endings, and sequences can be appealing to audiences, these stories may inaccurately portray the current state of scientific information on certain topics, especially ones for which much of the science is future-oriented and still developing. Science communicators and storytellers can use the information in this chapter to scale the setting of their stories more narrowly or more broadly and adjust the ordering of their narrative to engage audiences in their scientific topics.

Chapter 4 examines the tensions around storyteller and content. The storyteller is the source and/or sharer of the story who confers credibility on the story's content. Mistrust of the storyteller can thus corrupt the coherence and rationality of the story and its likelihood of being adopted. Scientific stories may challenge religious, political, or economic values by proposing evidence for alternative interpretations and understandings. The content of scientific stories may therefore disrupt people's worldviews and lead to skepticism or hesitation in the face of more convincing or resonating rival stories. Science communicators and storytellers can use the information in this chapter to reflect on who sciences' storytellers are, who rival storytellers are, how content will resonate with audiences, and how to adjust storytellers and content to increase credibility and relate to audiences' values and priorities.

Chapter 5 focuses on rhetorical strategies in science communication and what insights can be gleaned from the analysis of scientific controversies as competing stories. Because no one type of story is universally appealing, in lieu of "best practices," I use the term *constellation practices* to provide contextual advice, considerations, and questions for telling and evaluating scientific stories. Depending on one's audience, topic, context, and purpose, a science storyteller can adjust a story's narrative features. Learning from rival stories means locating strategic intersections and shared ground from a variety of scientific communities, as well as from religion, politics, and economics, as resources for productive coalition building. These intersections offer opportunities for rewriting the narrative of rivalry and competition into one of cooperation and mutual benefit.

In the concluding chapter, I briefly summarize the six narrative features and how they can be mapped using narrative webs. I also suggest how to navigate structural obstacles that prevent storytelling, such as those involved in research funding, publishing, media gatekeeping, and language practices. It is not only the actions of a single storyteller but also the institutions we find ourselves in that shape science's stories. The conclusion proposes future directions for inquiry and discusses the limitations and risks inherent in all communication, including scientific storytelling.

We must, perhaps now more than ever, not only evaluate potential reasons for scientific skepticism but also advocate for ways to improve science communication that are sensitive to audiences and deploy a variety of tools in thoughtful, critical ways. I hope this project leaves audiences and science communicators of all varieties with a rhetorical perspective on the pervasiveness of narratives and their value as a science communication tool. By embracing the narrative features inherent in science communication, we can empower science communicators and create more diverse, tailored, and purposeful stories.

I make no claims that storytelling will fully solve existing or future challenges to science communication. However, I do argue that storytelling is a useful tool for science communicators to capture more people who are otherwise skeptical of science or those who are confused by the circulation of competing narratives. By empowering science communicators to make use of narrative strategies, we can meet the public where they are and bridge what rhetorician Jeanne Fahnestock calls "the enormous gap between the public's right to know and the public's ability to understand" scientific information.[92] Together, we can transform the conflict of science *versus* story into the harmony of science *and* story. Critical attention to narrative features and the rhetorical nuances of audience, context, and purpose can help us all become science storytellers whose narratives resonate with people about the world and our place in it.

1 Case Studies and Rival Stories

In 1999, Climate Research Unit director Phil Jones sent an email briefly summarizing his process for combining measurements from climate proxy points, including ice cores and tree rings, to estimate global warming temperatures. Jones was concerned that the tree ring data were unreliable because they showed less warming than other measurements. By combining the tree ring data with more proxy measures, he was able to create a more accurate picture than from individual data sources alone. Jones's description of the complex process of data aggregation was cherry-picked from a series of more than one thousand emails that were hacked from Climate Research Unit servers and released to the public in 2009. Jones's email read, "I've just completed Mike's Nature trick of adding in the real temps to each series for the last 20 years (i.e., from 1981 onwards) and from 1961 for Keith's to hide the decline."[1] While the statement appears damning out of context, with the knowledge that data aggregation is a standard analysis practice and that the "decline" refers to concerns about the quality of certain data, the statement is more clumsy than manipulative. Climate skeptics took advantage of the leak of technical discourse by latching onto the contextless statements and using them as evidence that not only were climate scientists engaging in nefarious manipulation of data, but also that the emails supplied evidence that all of climate science is unreliable and fabricated.[2]

Also known as "Climategate," the Climate Research Unit's hacked emails are an example of how technical discourse can be interpreted in ways contrary to the intended meaning. Taken out of context, statements like "trick" and "hide the decline" appear to reveal deception perpetrated by climate scientists.[3] Although these were shorthand statements to communicate to members of the technical community familiar with these methods how Jones was working with the data, they appear to reveal ulterior motives. As communication scholar John Lyne and biologist Henry F. Howe argued, "When scientific talk catches the ear of the laity, the experts may completely lose control of the use of their expertise."[4]

Differences in how technical experts and members of the public use language can be explained through sphere theory.[5] Spheres are arenas of communication that involve different participants, appropriate reasoning, argumentative resources, and communication effects. The three spheres are the technical sphere (i.e., communities of experts), the public sphere (i.e., general communities), and the private sphere (i.e., interpersonal relationships and networks). Those familiar with the jargon, techniques, and background knowledge of the technical sphere of climate science were able to interpret Jones's statements as appropriate data aggregation methods to reach the most accurate conclusions about the state of global temperatures. Indeed, multiple independent groups cleared the Climate Research Unit team of any data manipulation or wrongdoing.[6] The technical sphere story about Climategate became one of statements taken out of context and overblown claims of foul play.[7]

However, the public sphere story became one of deceit, manipulation, and fudged data that put the Climate Research Unit and climate science conclusions into question.[8] Although it is hard to trace specific consequences from Climategate, some attribute the controversy to the "delay [of] measures that might have slowed climate change and given humanity more time to cut atmospheric carbon dioxide levels."[9] In other words, Climategate inserted enough doubt into public and political minds to halt action, despite being of sound technical and scientific practice.

Climategate is not alone in being a scientific controversy emerging from public-technical miscommunication. The controversy over the relationship between vaccines and autism was largely attributed to a single now-retracted article in *The Lancet* by Andrew Wakefield. In a public version of the narrative forwarded by vaccine skeptics, Wakefield is a heroic whistleblower who successfully challenged the consensus around vaccines in a peer-reviewed publication and was wrongly stripped of his medical license.[10] Alternatively, the technical version of the narrative paints a picture of small sample sizes, conflicts of interest, and inappropriate use of causal language to assert a link between vaccines and autism without sufficient evidence.[11]

Walter Fisher used the term "rival stories" to refer to stories that are largely incompatible with one another and compete for audience adherence. These narratives make up the discourse of scientific controversy by solidifying opposing sides, interpretations, and arguments. Rival stories, therefore, "deny each other in respect to self-conceptions and the world,"[12] because they create such different portrayals of audiences. In the case of Climategate, audiences are either justified in trusting climate scientists or they have been duped, based on the story one tells. I categorize rival stories that are harmful to the public good; are motivated by goals other than seeking, verifying, and sharing knowledge; and contribute nonproductively to public discussion of scientific topics as "disingenuous" rival stories.[13]

These stories might spread misinformation or disinformation, muddy the waters, or purposefully cast doubt on scientific conclusions for monetary or political reasons. Historians Naomi Oreskes and Erik M. Conway referred to groups and individuals who peddle uncertainty and misinterpretations of scientific evidence as "merchants of doubt." Instead of proposing their own evidence, merchants of doubt, or disingenuous rival storytellers, seek to distract, delay, and sow doubt to undermine scientific evidence that threatens "their commercial interests or ideological beliefs."[14] Disingenuous rival stories may be spread by those who tell them knowing they are misrepresentations, but also by those convinced of the story and hoping to spread it to enlighten others.

Not all rival stories are unproductive or disingenuous. There are ways to challenge and discuss proper courses of action regarding scientific conclusions that strengthen the scientific community and its practices. When we issue these challenges, they can be done in productive ways that build, expand, and collaborate with Eurocentric scientific knowledge to identify important gaps and build toward more diverse and inclusive knowledges. These "productive" rival stories elucidate voices that would otherwise be concealed, silenced, or forgotten, such as those told by marginalized communities.[15]

Humanities scholar Aja Martinez calls these stories, especially those that challenge dominant racial hierarchies, "counterstories."[16] Drawing on the work of Richard Delgado, Martinez notes that those in power "craft stock stories to establish a shared sense of identity, reality, and naturalization" that legitimize their "superior position" and normalize a particular story to the detriment of alternative stories.[17] Eurocentric science stories can be seen as "stock stories" in that dominant ways of knowing have legitimized their methods and approaches as "neutral" and reflected in "reality."[18] Similar to the demarcation practices discussed in the introduction chapter, stock stories frame stories of the "outgroup" as "biased, self-interested, and ultimately not credible," thereby excluding them.[19] Like racial counterstories, productive rival stories in science "serve the purpose of exposing stereotypes, expressing arguments against injustice, and offering additional truths through narrating authors' lived experiences."[20] These stories seek to challenge Eurocentric, dominant stories but with the goal of correcting, building, and reimagining scientific inquiry to be more inclusive and representative.

When faced with rival stories, people may choose to adopt one or the other, or create strategic blends of the two. For issues of human origins, audiences may have to grapple with whether they are products of natural selection (i.e., evolution) or divine intervention (i.e., creationism, Indigenous creation stories). Intelligent design advocates, however, leave room for supernatural or divine involvement in human origins while retaining the mechanism of natural selection. Tendencies for binary thinking may create overly simplified descriptions of scientific controversies as one side versus another. In many instances, there are competing narratives from a variety of sources that establish different audiences, values, and priorities.

In what follows, I chart brief overviews of the four main controversies discussed in subsequent chapters. These outlines lay the foundation for understanding the narrative features used in technical storytelling and rival storytelling, both disingenuous and productive, around scientific topics. These comparisons provide context for the case studies as well as introduce competing sources of information that public audiences weigh in decision-making.

A BRIEF OVERVIEW OF THE CLIMATE CHANGE CONTROVERSY

The burning of fossil fuels and the subsequent release of carbon dioxide into the atmosphere is causing major disruptions to Earth's climate. The interconnected climatic changes have been referred to as global warming, climate change, climate crises, and climate emergency. Whatever we call it, a warming planet threatens life as it currently exists; unmitigated consequences will be "catastrophic and manifold."[21] Intergovernmental Panel on Climate Change scientists predict that rising temperatures will lead to more intense and frequent weather events, negative impacts to human health, and damages to ecosystems worldwide (reported from medium to high to very high confidence levels).[22]

Scientific consensus has repeatedly established that human actions have negatively impacted the climate,[23] causing a new era of geological time: the Anthropocene. Despite historical apathy and polarization on the topic of climate change, recent polls show positive changes in belief and concern about climate change.[24] This is reason for optimism, but roadblocks to comprehensive political action remain. For example, only a slight majority of Americans (62 percent) believe that global warming is caused by humans.[25] Additionally, six in ten Americans say they "rarely" or "never" discuss climate change with friends and family, and less than half report hearing about climate change in the media at least once a month.[26] There is scientific consensus on climate change's existence and its threats to humanity, but this technical certainty is not reflected in policy action, which is frequently halted by a variety of delay tactics.[27]

Climate skeptical rival stories come from a variety of sources, including political, economic, and religious storytellers, that complicate adherence to climate science stories. In subsequent chapters, I focus on two disingenuous rival stories, which I call "Christian fundamentalism" and "neoliberalism," which are religious and economic stories, respectively. These rival stories are often intertwined with politics and support one another in raising doubts about climate science. For example, there are links between conservative politics, economics, and positions on social issues that attract Christian and primarily Evangelical voters.[28] US liberal voters tend to be more concerned about climate change than conservative ones, although this is not a strictly deterministic relationship.[29]

Religious identity can be an important factor in many people's climate change beliefs.[30] Many faiths directly support environmental protection, such as Judaism, Islam, and Hinduism. Others, such as Christianity, have a more tepid relationship.[31] Religious resistance to climate policies tends to be rooted in a variety of beliefs. For some, God rules over nature, meaning that any environmental changes are not due to human behavior, but to God's plan.[32] Some Christians see changing climates as evidence of the Second Coming, which should be either welcomed or considered as inevitable.[33]

In the disingenuous rival stories of neoliberalism, conservative non-profits often emphasize protecting human life over the environment by advocating for free market economic and energy policies. Neoliberalism values people as individual, economic actors, distrusts collectivism and economic sacrifice, and positions profits as the only rational decision-making choice.[34] Neoliberalism's stories shift attention from the environment to the economy and appeals to the values of American individualism and capitalism.[35]

Part of the reluctance to address the precarity of our environment can be found in the persistent Western ideology that humans are superior to nature, enabling the exploitation of the environment.[36] Nature has "been displaced in numerous narratives, including Christian, Enlightenment, scientific, capitalistic, socialistic, and industrial" ones in favor of "plac[ing] human reason and humans at the center."[37] Indigenous scientist Jessica Hernandez (Binnizá & Maya Ch'orti') notes that the separation non-Indigenous systems create between humans and our environment "prevents us from seeing them as our relatives" and instead puts a "price tag" on their exploitation and utility.[38] The persistent notion of nature as divided from humanity, including how we often express this relationship through language, reinforces the idea that humans are one thing and the environment is a separate thing altogether.[39]

Studies repeatedly show, however, that climate change will affect everyone, and some of us more than others. Namely, marginalized communities, including women, communities of color, and those living in poverty, will be most adversely affected by our changing climate.[40] Productive rival stories, such as those forwarded by marginalized communities, can point out areas scientists and science communicators have overlooked or failed to incorporate. Rhetoricians Steven Schwarze and colleagues noted that scientific and technical discourses "dominate public discourse and crowd out discussions of equity and social justice."[41]

Indigenous rival stories to Eurocentric science emphasize this interconnectedness and the already real and felt consequences of climate change. Indigenous and American Indian Studies scholar Daniel Wildcat (Yuchi, Muscogee) argues that Indigenous communities have already undergone three "removals," including geographic removal from the land, and social and psycho-cultural removals that displace Indigenous ways of knowing and their right to sovereignty and self-identity.[42] Indigenous science stories challenge Western academic notions of knowledge and disciplines to embrace a "holistic system" that forwards integrated and interconnected understandings of human-nature relationships.[43]

Additional productive rival storytellers include youth protestors at COP meetings who argue that leaders are not working fast enough or with enough urgency to forestall environmental damages. Specifically, African youth activists, including Elizabeth Wathuti, Hounaidat Abdouroihamane, and Vanessa Nakate, have pointed out that COP meetings repeatedly ignore and underplay the needs of Africans and how the African population is suffering disproportionately due to climate change and climate inaction.[44] These narratives challenge dominant notions that COP meetings are effective or are at least addressing the full picture of climate justice. In an illustrative example of Wildcat's notion of "removal," Nakate was cropped from an image posted by the Associated Press documenting a climate conference at Davos.[45] Instead of the five youth activists being represented, Nakate's removal left only four White youth activists in the image.

Productive rival stories challenge technical stories about climate change but do so with the hope of transforming human-nature relationships and accelerating climate action by including more voices and shifting harmful, hierarchical relationships into holistic and collaborative ones. Through counterstories, missing, excluded, and undervalued voices and perspectives can be included and amplified to create more equitable and comprehensive environmental deliberations.

A BRIEF OVERVIEW OF THE EVOLUTION CONTROVERSY

Around 3.7 billion years ago, microscopic organisms left evidence of the first life on Earth. As these organisms began living inside of one another, multicellular life emerged. The first members of the animal kingdom, sponges in the depths of the ocean, evolved around 800 million years ago. Through gene mutation, which branched life into different features, the entirety of Earth's lifeforms developed, died off, reproduced, and adapted. This complex and incredibly long process led to the evolution of humanity's ancestors, such as Lucy and Ardi, into what we would recognize as modern humans, *homo sapiens*.[46] The theory of evolution is the accepted explanation for human origins because it is our best operating explanation for life's diversity and complexity.[47] Although there is still much to research about the process, scientists have confirmed no evidence contradicting the foundational tenets of evolutionary science.

Despite the scientific consensus, skepticism and opposition persist. Evolution's disingenuous rival stories are creationism and intelligent design, which emerge as religious, primarily Christian, challenges to some of evolution's tenets. These stories function to undermine scientific knowledge and sow mistrust of scientific processes. Indeed, the controversy over evolution "has grown in both size and intensity" in recent years[48] and "has persisted for an extraordinarily long time"[49] despite its acceptance in scientific thought and public education. Charles Darwin's *On the Origin of Species* was a foundational text in evolutionary science that collected evidence in support for the model of what he called "natural selection, as the primary driver of evolution."[50] Darwin's theories amended and, for some, replaced Biblical creation stories, although he did not specifically exclude the possibility of the divine. He even called the process natural *selection* to imply "foresight, intelligent planning, and craftsmanly skill" in the process.[51] However, his theories did shrink the role that divinity played and subsequently complicated literal interpretations of the Bible. Some interpreted his conclusions as disproving religious conclusions, or at least dealing them a serious blow.

After Darwin's research, additional experiments and material evidence supported natural selection and evolution. In the twentieth century, the human origins controversy moved into the schools, where creationism previously had been central.[52] Public schools became the primary "battleground" for the morality of the youth.[53] Some equated evolutionary science in classrooms with the removal of religion from the public sphere. Throughout the 1900s, there was a series of court cases that solidified evolution's place in science classrooms.[54] These cases, including a few in the 1990s and early 2000s,[55] upheld a standard of separation in which science and religion were distinct fields of inquiry, and creationism, being religious, was decidedly *not* scientific, and thus had no place in public classrooms.

Recent iterations of the evolution controversy have extended beyond the classroom. Communication scholars Casey Ryan Kelly and Kristen Hoerl argue that nonprofit creationism advocacy groups are becoming more prominent, noting that "creationists have sought out other venues [besides the courts] in which they might establish themselves as legitimate stakeholders in a public controversy over scientific theory."[56] Creation museums, for example, are tourist attractions that draw thousands of visitors annually and challenge evolutionary science through offering competing narratives in a familiar form.[57] Creation museums exist in the United States and across the globe, from Kentucky to California and from England to Hong Kong.[58]

While the United States is a hotspot for creationism, polls consistently show that evolution skepticism exists in countries across the Middle East, Latin America, and Europe.[59] Even in countries with high acceptance of evolution, there are sizable pockets of evolution skeptics—up to a quarter or a third of the population.[60] In a multinational survey conducted in 2020, 74 percent of adults responded that humans have "evolved over time."[61] US respondents polled below that figure (65 percent), with most of the variation being attributed to religious affiliation, primarily believers in Christianity and Islam.[62] Although there is evidence that belief in evolution is increasing, there remains a thirty-three-point gap in the acceptance of evolution between the public and scientific communities.[63] This gap, however, cannot be attributed to simple knowledge deficits. The National Science Board reports that nearly three-quarters of the people in the United States (72 percent) show knowledge of evolution and its components, even if they do not believe evolution is true.[64]

The human origins controversy reinforces the binary of science (i.e., evolutionary science) and religion (i.e., creationism) as mutually exclusive epistemologies. Intelligent design has emerged as a third narrative that includes some aspects of science and marries them with religious belief. Intelligent design's stories share the divine creator of creationism's stories, but leave room for the processes of natural selection as the means by which God achieved the variety of life seen today, including human life. However, this position often draws disdain from both evolution and creationist adherents.[65]

Although the war frame of science versus religion may be appealing and motivating to some, to others it is simply not an accurate representation of their faith. Unlike fundamentalist Christianity and Evangelicalism, other religious and Indigenous creation stories exist alongside evolutionary science. Many faiths, even those who have their own creation stories, view their faith as compatible with evolutionary science, thereby posing a counterstory to "science versus religion." Pope John Paul II, for example, noted that evolution has become the reigning explanation for human origins in a material sense, while still noting that "the spiritual soul is immediately created by God," creating room for evolutionary science and the Catholic faith to coexist.[66]

Similarly, Indigenous creation stories voice specific community beliefs about tribal origins that rarely are posed in opposition to evolutionary science. Hernandez notes that in the creation story for Zapotec people, "our ancestors were born from trees and jaguars. . . . We came from earth, and this explains the strong connection we have with our Mother Earth."[67] Multidisciplinary practitioners Zahraa Saiyed and Paul D. Irwin argued that Indigenous creation stories explain "the relationship between humans and nature" and foster a sense of care for the Earth, "not only [as] a sustainer of life, but as the original source of all life."[68] Education scholar Norbert Francis argued that perceived "contradictions" between Native American creation myths and evolutionary science are largely "without foundation."[69] Instead, these stories provide cultural, philosophical, and religious guidance that can be compatible with the science of human evolution.

A BRIEF OVERVIEW OF THE VACCINATION CONTROVERSY

Vaccines are medical substances typically administered through injections to people at a variety of ages to defend them against contracting diseases by activating the body's immune response. Many diseases kill humans before the body has a chance to fight off the microorganisms causing them, but by controlling the body's exposure to a weakened or dead form of the disease, the body can build up protection before live contact. Vaccination has functionally eradicated many diseases but has faced perennial opposition since it became widely adopted in Europe in the 1800s.[70] Edward Jenner developed what is considered the first vaccine in 1796 when he demonstrated that exposure to cowpox provided immunity against the smallpox virus. The Vaccination Acts of 1853 and 1867 mandated vaccinations for children and imposed a fine or jail time for failing to do so. Medical researchers Robert Wolfe and Lisa Sharp described these laws as "political innovation[s] that extended government powers into areas of traditional civil liberties in the name of public health."[71] Consequently, anti-vaccination groups formed in their wake, primarily concerned with the compulsory elements of these laws and concerns over vaccine safety.[72]

Many early concerns regarding vaccinations live on in contemporary vaccine hesitancy. For example, in a content analysis of anti-vaccination websites, medical anthropologist Anna Kata found arguments related to safety, effectiveness, and civil liberties, among others, as frequent themes.[73] There are many reasons why people may be vaccine hesitant, such as skepticism of government influence, fears about vaccine ingredients, and sincere religious objections.[74] A US Gallup poll conducted in early 2020 found that there has been a 10 percent decline in survey respondents agreeing that vaccinating children is important since 2015, and that 56 percent of Americans either believe that vaccines cause autism or are unsure if they do.[75] The rise in vaccine skepticism, no matter the legitimacy or reasonableness of the decision, has been linked to outbreaks of previously eradicated diseases, such as measles, in the United States.[76] The World Health Organization lists vaccine hesitancy as one of the top ten threats to global health, due to its contribution to declining vaccination rates.[77] It is important to note that vaccine hesitancy and vaccine exemptions of all types pose potential public health risks by delaying vaccine uptake to reach herd immunity.[78]

The contemporary iteration of the vaccination controversy arguably began with the 1998 publication of a *Lancet* article by Wakefield and colleagues that proposed a link between vaccines and autism. Due to a variety of flaws, the article was retracted in 2010 and Wakefield had his medical license suspended.[79] But, this retraction and suspension has not stopped individuals from coming forward in support of a vaccine-autism link. Indeed, Wakefield's discrediting is often mentioned on anti-vaccination websites as evidence of a larger conspiracy to cover up the vaccine-autism link.[80] While many vaccine skeptics are worried about the relationship between vaccines and autism, historian of science Mark A. Largent notes that focusing solely on the (in)accuracy of autism claims in vaccine discourse "obscures serious problems" that parents also see with the current vaccine schedule and overly narrows the conversation to technical concerns.[81]

Medical information is not solely constructed in doctors' offices and by institutions such as the Centers for Disease Control. Public constructions of diseases, disorders, and treatments emerge in media, politics, and personal networks, which can work with but also against institutionalized scientific conclusions. Many arguments about vaccines have endured over time, but the methods of sharing vaccine hesitant messages have changed. The internet has become a fertile ground for the proliferation of vaccine skepticism. Websites and social media communities share "vivid personal narratives" with high emotionality that drive vaccine hesitant parents to consume and further share those stories, giving people the impression that vaccine injuries are a widespread and legitimate fear.[82] Online, people can easily find a bevy of resources, including vaccine exemption materials.[83]

Rival stories of vaccination tend to empower audiences to make decisions for themselves and center family health over community well-being. I call the first the narrative of "individualism" and the second the narrative of "vaccination conspiracy." The stories of individualism empower parents to make decisions for their families without government intervention. This narrative proposes that mandating any aspect of health-related issues violates personal autonomy. Despite the prevalence of national health and safety regulations, such as federal limits on smoking and not wearing seat belts, appeals to personal liberties have been a mainstay throughout anti-vaccination's long history.[84]

Scholars attribute these differences in part to the invasive, personal nature of vaccination and its perceived irreversible effects.[85] Writing and rhetoric scholar Heidi Yoston Lawrence argues that part of the decision-making process in vaccination comes down to the notion of the "irreparable."[86] By the irreparable, Lawrence means that the decision to vaccinate is an irreversible one; there is no going back to an unvaccinated body once vaccinated. Similarly, once exposed to a harmful disease one is not vaccinated for, one cannot return to an unexposed body. With competing stories circulating both on- and offline, vaccination decision-making becomes especially fraught. Parents want to make the best and safest choices for their families, so must weigh increasingly large amounts of information, risks, and possible futures with consequences that cannot be undone. Vaccine hesitancy or delay, or forgoing certain shots, may be viewed by parents as reasonable ways to manage these fears that are under their control.[87]

The rival stories of vaccination conspiracy detail an elaborate collusion of government, doctors, and "Big Pharma" (i.e., the pharmaceutical industry) that seeks to harm the population, oftentimes children. Similar to the individualism rival narrative, the vaccination conspiracy narrative provides parents an important role to play in their children's health. This narrative characterizes doctors as cruel and malicious, and thus, untrustworthy. A narrative of evil conspiracy locates parents in an important, if not the most important, role in the story: savior.[88] Parents are empowered to protect their children against a formidable enemy of elite and powerful individuals and institutions.

Productive rival stories to vaccination prompt recognition of disparities in medical access and the long history of medical malpractices, especially levied on marginalized populations. In *The Origins of Bioethics: Remembering When Medicine Went Wrong*, rhetorician of science John Lynch highlights important case studies of medical injustices where studies were run on vulnerable communities without consent or full information regarding risks, or treatment, sometimes even purposefully infecting them with harmful pathogens.[89] Additional examples of medical injustices include unauthorized use of Henrietta Lacks's cancer cells or the over sixty thousand forced sterilizations of "undesirables" such as people convicted of crimes, mental health patients, people of color, disabled people, and immigrants in the United States in the 1900s.[90] These incidents are important to identify, acknowledge, and remember to uphold ethical medical practices and support the public good. Even as we sympathize with the complexity of these decisions and recognize the ways that medical systems have not always served the public good, it is important to note that vaccine hesitancy poses serious threats to public health. These risks are increasing as the probability of novel viruses and pandemics, such as COVID-19, become more likely with rising global temperatures.[91]

A BRIEF OVERVIEW OF THE COVID-19 CONTROVERSY

In December 2019, the Chinese government alerted the World Health Organization about increased reports of people being hospitalized with "severe pneumonia."[92] While the origins of COVID-19 is still a topic of inquiry, there is considerable support that the earliest cases emerged from a series of "zoonotic events," where diseases spread from animals to humans, at the Huanan Seafood Wholesale Market in Wuhan.[93] Soon thereafter, the disease spread throughout Asia and had reached Europe and the United States by late January 2020.[94] COVID-19 is caused by the SARS-CoV-2 virus and manifests in the body with a variety of symptoms, including dry cough, fever, fatigue, shortness of breath, loss of taste or smell, headaches, chest pressure, joint and muscle pains, skin rashes, and confusion (i.e., COVID fog).[95] In more severe cases of COVID-19, people experience additional symptoms such as myocarditis (i.e., heart inflammation), brain inflammation, and strokes.[96]

On March 13, 2020, then–US president Donald Trump declared COVID-19 a national emergency, which provides the federal government additional powers to aid in rapid response, including additional funds for mask production and vaccine research.[97] After the national emergency announcement, Trump instituted travel restrictions and states began implementing stay-at-home orders, which would last for months. During these measures, nearly every state capitol at some point witnessed swarms of unmasked protesters—many of whom bore firearms—demonstrating against lockdown orders or mask mandates. These decisions came from a belief that injunctions to stay home were violations of individuals and their freedom.

By December 2020, COVID-19 vaccines developed by Pfizer and Moderna were approved by the Federal Drug Administration. Sandra Lindsey, an ICU nurse in New York, was the first to receive a COVID-19 vaccination on December 14, 2020. Vaccinations were first given to health care workers, including workers and residents in long-term care facilities. As vaccine rollouts continued, increasing numbers of people qualified to receive COVID-19 vaccines. As of April 2023, 81 percent of the US population and 72.3 percent of people worldwide have received at least one dose of a COVID-19 vaccine.[98] While these rates are encouraging, they are far from the predicted 94 percent immunity rate needed "to interrupt the chain of transmission" of COVID-19.[99] These statistics also hide the disproportionate distribution of vaccines globally; areas of the world with the lowest percentage of their populations vaccinated are Africa and the Middle East and only 22.3 percent of people in low-income countries have received one or more doses of COVID-19 vaccines.[100] The COVID-19 controversy is, in part, related to growing vaccine hesitancy. Polls early into the vaccine development showed that less than half of Americans (44 percent) reported that they would get a COVID-19 vaccine.[101] These numbers are improving, in part due to people gaining confidence in the COVID-19 vaccines as more people are getting vaccinated with no adverse reactions.[102]

The disingenuous rival stories to COVID-19 are similar to those of vaccination. Many people saw preventative mandates such as mask wearing, stay-at-home orders, and vaccination as infringements on individual liberties. There are also similarities in the circulation of conspiracies regarding various aspects of COVID-19 to compete with official narratives from figures such as the Centers for Disease Control and Anthony Fauci. Conspiracies about COVID-19 encompassed myths of COVID-19 origins, the contents of the vaccines, and the purpose of mask wearing.

I call the first disingenuous rival narrative to COVID-19 "freedom over fear." This narrative positioned a dichotomy between preventative measures (e.g., lockdowns and masking) and personal freedoms. Those who participated in preventative measures were portrayed as fearful of the virus and as giving up their rights and freedoms to the government. Co-opting the "my body, my choice" language of abortion access advocates, some opponents to COVID measures emphasized personal choice over mandates. In their examination of conspiracies and misinformation circulating online during the COVID-19 pandemic, communication scholars Dror Walter, Yotam Ophir, and Hui Ye located a similar theme they called "containment," which was discourse that associated public health measures such as masking and lockdown mandates as "infringing liberty and freedoms."[103] In particular, some protesters advocated for freedoms *to* in addition to freedoms *from*. For example, lockdown protesters wanted the freedom to work and opposed lockdowns that closed down many businesses.[104] For people who find a sense of identity in their work and/or are opposed to what they perceive to be government handouts, lockdowns were a serious threat to their values.[105] Others opposed extensive lockdown measures because of limits placed on daily life.

The second disingenuous rival narrative to COVID-19 stories is "COVID conspiracy." Similar to vaccination conspiracies, COVID-19 conspiracies characterize scientific conclusions as motivated by something other than medical evidence, such as financial gain. Many COVID conspiracies questioned the source of the SARS-CoV-2 virus as purposefully created and released to amplify intergroup conflict.[106] For example, some conspiracies portrayed the virus as a bioweapon and thereby perpetuated Orientalist and racist tropes against Chinese and Asian people.[107] Other conspiracies amplified political divides and considered the virus to be a political tool between liberals and conservatives in the United States.[108] The spread of these conspiracies, among others, damages public health because belief in COVID-19 conspiracies is associated with "unwillingness to social distance and vaccinate against the virus."[109]

These disingenuous rival stories are only part of the narrative discourse around COVID-19. Similar to productive rival stories in the vaccination controversy, COVID-19 productive rival stories recognize that low vaccination rates, especially in the United States, were not entirely attributable to vaccine hesitancy. For some communities, vaccine rollouts were limited and accessing these sites was incredibly difficult. Communities of color had low vaccination rates in part due to long histories of harmful medical interventions, but also due to already impoverished health care systems now additionally responsible for distributing vaccine resources.[110] The Centers for Disease Control reports that Black and Latino people are hospitalized due to COVID-19 at five times and four times the rate, respectively, of non-Hispanic White people.[111] When communities of color are treated for COVID-19, issues of racism in the health care industry also emerged. One illustrative example is the tragic story of Dr. Susan Moore, a Black woman and physician who was denied pain medication for shortness of breath and had her concerns ignored after being admitted to the hospital after contracting COVID-19.[112] Moore's death was labeled by many as an act of medical racism as she recorded her (mis)treatment by White doctors.[113] Productive rival stories point out discriminatory gaps in COVID-19 responses that leave communities of color and low-income communities behind in prevention and treatment while attributing the bulk of the issue to hesitancy.

The COVID-19 narrative is still unfolding at the time of writing as new mutations emerge, restrictions and personal protections have been mostly abandoned, and the effects of "long COVID" are still being discovered. While the COVID-19 case study has many similarities with that of vaccination, the unique features of masking, lockdowns, and the experience of living through a pandemic have shaped public and scientific narratives on the topic in distinct ways.

SIMILARITIES ACROSS RIVAL STORIES

Scientific inquiry is concerned with knowledge development and is continually in the process of growing, clarifying, refining, and experimenting, which creates new knowledge, upends old knowledge, and revises existing recommendations. Challenging mainstream conclusions is thus integral to how scientific knowledge develops through debate, clarification, and revolution.[114] Productive rival narratives challenge institutionalized science by reinserting missing or overlooked voices, correcting misunderstandings, and making space for alternative knowledges and histories. In the case studies above, I have highlighted examples from Indigenous communities, communities of color, and religious communities whose counterstories expand scientific understandings of climate risks and consequences, creation stories, and health disparities to be more inclusive and representative.

Disingenuous rival stories, however, deny, stall, and oppose scientific stories in unproductive and harmful ways. Disingenuous rivals cherry-pick alternative authorities who serve as credentialed outliers that question scientific consensus. In the case of climate change, climate deniers appeal to a minority of scientists who doubt the scientific consensus and propose caution and additional research before taking action toward mitigation and adaptation.[115] In the case of evolution, creation and intelligent design advocates run their own research centers and propose alternative interpretations of scientific facts to challenge evolutionary science.[116] In the case of vaccines, defectors from the medical community leverage their medical credentials to speak about the harms of vaccination.[117] For COVID-19, the authority of the Centers for Disease Control was put into question as policies and health advice shifted, especially early in the pandemic.[118] Skeptics leverage alternative authorities to cast doubt on scientific conclusions and to paint the scientific community as fractured and uncertain.[119] These authorities may be political or religious leaders, economic experts, social media communities, and technical outliers.

Another important similarity is the presence of powerful, well-funded groups that support disingenuous rival stories. These interested parties often have a stake in keeping the controversy alive, so they actively foment dissent, uncertainty, and doubt. The climate change "denial machine," made up of fossil fuel industries, conservative think tanks, and fringe media outlets, downplays the severity of climate change for monetary and political gain.[120] In the case of evolution, creationist and intelligent design organizations receive millions in private donations to build and maintain creationist tourist attractions, run home school programs, and engage in public debates.[121] While much anti-vaccine hesitancy seems to stem from concerned parents, there are prominent examples of anti-vaccine advocates who profit by selling vaccine alternatives.[122] For COVID-19, "Republican megadonors" funded the Convention of States project that organized dozens of rallies across the United States aimed at protesting lockdown measures and supporting "re-opening" the economy.[123]

Rival stories, their similarities, and their differences offer promising comparison points to begin experimentation and exploration with narrative features that may be more engaging for and appealing to audiences. From disingenuous ones, we can learn about the persuasiveness and pervasiveness of them as narrative structures. For example, the disingenuous rival stories of creationism might provide insight that the content of evolution's stories is not resonating with certain audiences, necessitating an acknowledgment of these values and worldviews or location of strategic overlaps. The disingenuous rival stories of individualism in the vaccination case study can inform our understanding of incongruence between character and action where guardians feel displaced from health care decision-making.

From productive rival stories, we can learn to break down barriers and redraw the lines of demarcation to reflect the "racialized, classed, and gendered" positionalities of stock and counterstory storytellers.[124] The productive rival stories of health disparities in vaccination could prompt a reconsideration of science's storytellers to those who are members of a marginalized group to infuse the story with trust and confidence. The productive rival stories of Indigenous science could encourage science communicators to narrow their scopes to community impacts and address environmental justice as a foundational component of climate change.

These patterns provide insight into narrative opportunities for science communicators. In the presence of persistent rival narratives, how can we improve upon scientific storytelling to engage audiences and garner additional support for scientific conclusions and their implicated policies? How can we learn from productive rival narratives to create more inclusive and equitable scientific practices, messages, and communities? Narrative webs and their wedges aid in our exploration of how science communicators can modify, experiment with, and improve their science communication through attending to rival stories' narrative features and reflecting on trends in science communication that may inhibit narrative engagement. While these case studies are not representative of all iterations of scientific controversies, they are important examples of some alternative stories currently circulating in science communication. With the provided context in mind, we can dive deeper into the specific discourses employed by a variety of storytellers in the subsequent analysis chapters that are organized by pairs of narrative wedges.

2 Character and Action Wedges

In late December 2020, a Wisconsin pharmacist, Steven R. Brandenburg, removed more than five hundred doses of Moderna's COVID-19 vaccine from their needed deep-freeze cold storage. He was charged with three felonies—reckless endangerment, drug tampering, and criminal property damage—and plead guilty to the charges a month later.[1] While his motives were initially unclear, the plea filing revealed that Brandenburg was "skeptical of vaccines in general and the Moderna vaccine specifically," so tampered with the doses to make them ineffective.[2] In late 2020, the vaccine rollout had only just begun, meaning that Brandenburg had tampered with vaccines going to frontline workers, like himself, and vulnerable elderly populations. For many, Brandenburg was a "bad actor" who caused "multifaceted and severe" harm to people injected with the compromised doses in the form of uncertainty and anxiety over the vaccine's effectiveness.[3] For others who were skeptical of the quick development of the COVID-19 vaccines, Brandenburg was a "hero" who validated conspiracy theories from inside the health profession.[4]

As a character in one story of COVID-19, Brandenburg can occupy the role of hero or villain depending on who is telling the story and to what audience. While there is little disagreement over the facts of Brandenburg's case, whether one considers his actions to be positive or negative changes the retellings' narrative implications. Much like CJ's story from the introduction, when Brandenburg's story, such as the version told by the *New York Times,* is mapped on a narrative web, many of the wedges are on the micro-ring. Brandenburg's story is one of a specific event with specific people in a particular time and place performing specific actions. The narrative specificity of examples may be, in part, why news stories are so engaging for audiences. Unlike CJ's story but typical of news articles, the coverage of Brandenburg emphasized his actions as part of the broader COVID-19 pandemic. For example, prosecutors in Brandenburg's sentencing argued that the judge must take into consideration how the "notoriety" of Brandenburg and his actions "could hurt the cause of persuading everyone to be vaccinated."[5] When an example is being used to introduce a topic in order to address more abstract concerns, we can map two distinct scopes on the web, where Brandenburg's actions are the vehicle between a smaller and larger scope (see figure 3).

Brandenburg's story is particularly attention grabbing due to the misalignment of character and actions. Brandenburg is a member of a technical community of medical professionals, meaning that his character might give the primary impression of being pro-vaccine, or, at least, not anti-vaccine. However, his actions show a disregard for his medical training by sabotaging vaccine doses and potentially faking his own vaccine status for employment.[6] Brandenburg may seem like an outlier in this regard, but the truth is much more shocking. Polls show that the rate of COVID-19 vaccine skepticism is higher in medical professions than in the general public, with some states reporting ranges of 40–60 percent of frontline workers, nursing home workers, and hospital workers unwilling to take the vaccine early in the vaccine rollout.[7] This surprising correlation has been attributed to the widespread misperceptions about the speed at which the vaccine was produced, whether more research is needed to assess the potential for side effects, loyalties to anti-vaccination politicians, and the circulation of vaccine myths on social media.[8]

Simply put, frontline workers and medical professionals are people; they are members of the public who may work in technical spaces and may have advanced training but are still subject to alternative messages about vaccines. The novel situation of COVID-19 has been particularly fertile

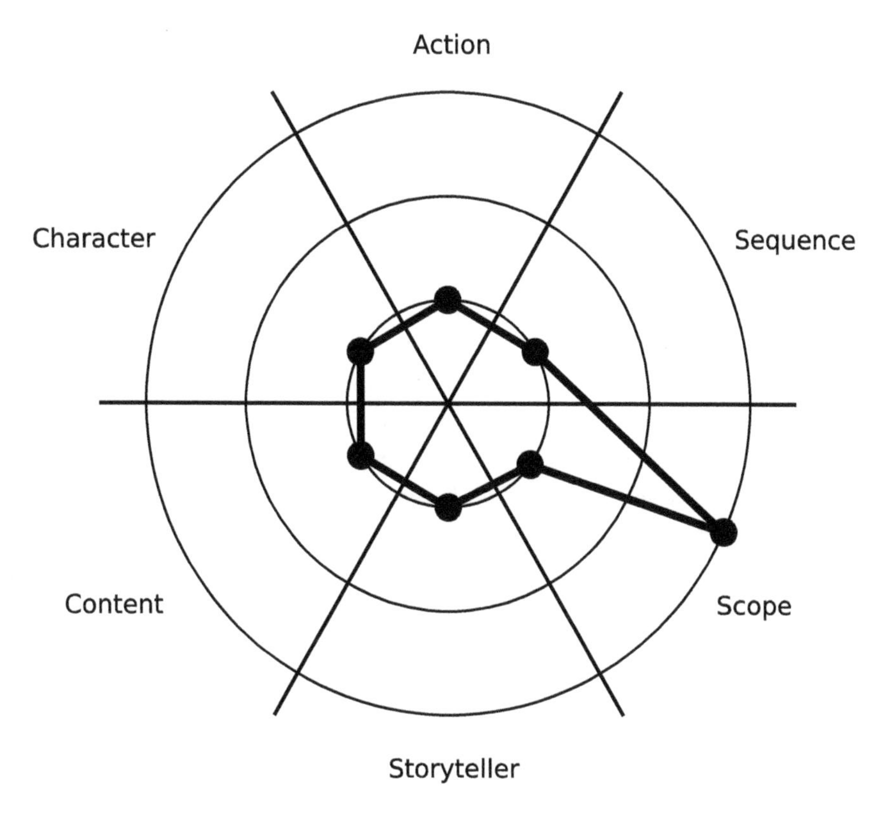

Figure 3. A narrative web of news coverage of Brandenburg's vaccine tampering showing two dots mapped on the scope wedge to reflect news stories' dual focus on the narrow and broad implications of Brandenburg's actions.

ground for the spread of myth, conspiracy theory, and politicization of vaccine prevention strategies. These factors all contribute to vaccine hesitancy and, consequently, a fluctuating number of people, in the medical profession and otherwise, willing to take the COVID-19 vaccine or other precautions. Each decision point is an opportunity for stories to take hold and to influence whether people choose to vaccinate, get a booster, or not.

This chapter explores the dynamics of characters and the actions they take in scientific stories. By viewing characters and actions together, we are focusing on the "who" and the "what" of stories: what happened and who made it happen or was involved in the happening. Rhetorical theorist Kenneth Burke calls these features the act and the agent of the pentad, the primary tool of his theory of dramatism. For Burke, acts and agents are co-constitutive. In other words, one can predict the kind of actions one might take based on the nature of the agent, and the nature of the agent is, in part, determined by their acts. For Brandenburg, the misalignment between act and agent may create a lack of narrative probability (how likely is it that a medical professional would intentionally ruin vaccines?) until one learns of the disturbing trend of anti-vaccine beliefs in medical fields, which provides the missing information to explain the character's actions and create a rational, believable story.

To analyze the dynamics of character and action and their varying implications for scientific stories, we must first dive into the wedges in more detail. Then, I provide examples from our four scientific controversies to examine how characters and actions have been constructed and may be further modified in scientific storytelling.

THE CHARACTER WEDGE

The character wedge is mapped at the micro-ring on the narrative web when a story's characters are humans similar to the audience and have clear, identifiable purposes. This may be a particular person being described in a news story, such as Brandenburg, a grandparent recounting their life experiences, or a main character on a hero's journey in one's favorite film or TV show.[9] The micro-ring, closest to the center of the web, would also include representative anecdotes, which are generalized parables.[10] For example, Abraham Lincoln's campaign slogan "Don't change horses midstream" is based on a fictional story of someone giving advice not to upset the current course by switching out horses.[11] While these individuals are fictional, they represent a type of conversation that could happen between two people. The micro-ring on the character wedge includes stories with a narrow, specific, and concrete character or set of characters who are human and are similar to the audience.

A story about "quasi-humans," such as anthropomorphized animals, "or other sentient beings" would be mapped in the center meso-ring.[12] For example, a fable about ants and grasshoppers is a story that has insect main characters, as opposed to human ones. While nonhuman animals and other beings are less similar to people, they are often anthropomorphized in stories to increase identification with them.[13] Storytelling across cultures incorporates anthropomorphized animals, such as Aesop's fables, African folklore of Anansi the Spider, or Southern Paiute origin stories. Nonhuman animals, plants, and spirits are frequent characters who are represented as communicating meaningfully in human and more-than-human languages.[14] These characters may be individual nonhuman animals, such as Peter Rabbit or Anansi the Spider, or may stand in for collective species, such as in Aesop's *The Ant and the Grasshopper*.

The meso-ring would also include collective human characters such as communities and groups. For example, a story of a protest may have many individual characters but would likely be expressed by discussing "the protest" or "the group" of protestors as a unit as the character of the story. Unlike the specificity of the micro-ring, the meso-ring creates more nebulous groupings that bring with them opportunities and obstacles. Discussing a group, for example, could provide evidence of widespread support for an issue whereas a story about an isolated individual might not. However, it may be harder for audiences to identify and find similarity with an entire group as opposed to a particular person.

A story would be mapped on the macro-ring when its characters are abstract, nonhuman, not readily identifiable, and lack purpose. A macro-ring story might have no discernable characters or focus more on "substance," "processes," and "materials."[15] Nonliving beings may be anthropomorphized or there may be little to no characterization present. For example, descriptions of atoms interacting with one another to form molecules or recipes that narrate the steps to create a baked good are largely limited to implied characters, combinations, and processes. The macro-ring is the least specific in terms of the characters presented and may appear to lack any characters at all. A Western, scientific story about the beginnings of the universe, for example, would lack any human-seeming characters in detailing the combustion and ionization of matter as the first lights turn on in space. Even the phrase "turn on" implies no purposeful agent or character behind various happenings in the story.

Some work in communication and rhetorical theory suggests the persuasiveness of taking a micro-ring character approach by using characters who are similar to the target audience, centering human experiences, and fostering connection between the story's characters and the audience. Indigenous methodologies scholar Shawn Wilson (Opaskwayak Cree) argued that everything, including academic research, is rooted in relationships, highlighting the need for connection between a story and its audiences.[16] Similarly, Burke argued that the most appealing stories let "the reader follow the story in terms of [a] single consciousness."[17] People respond to stories that have characters like them, with whom they can identify, form connections, and engage in the story.[18]

Even though focusing on individual people and the human element of storytelling can be effective, there are varying contexts, audiences, and purposes that necessitate attention. If we only aim to tell stories to general audiences, we may well create the impression that only one type of character is ideal for science's stories, such as offering a particular race, gender, and class that stands in as "neutral," which reifies hierarchies and creates exclusionary spaces in science. Additionally, micro-ring-character stories may unnecessarily limit who we consider to be agents capable of performing acts and ignore broader and more complex agents and characters at play in the stories we tell. For example, in telling the story of Brandenburg as an isolated character, we potentially limit our ability to address Brandenburg as one part of a larger population of vaccine skeptics who are also involved in online communities and groups related to anti-science conspiracy theories.

Meso-ring and macro-ring characters are not necessarily a sign of a weak story. The content of some stories simply lacks human characters or may operate on such broad scopes, such as the stories of evolution, where individuals are outside of the stories' purview or focus. Storytellers who wish to emphasize systems-level relationships may find it difficult or potentially compromising of the science to insert human characters, so may prefer to map their character wedge on the macro-ring. Individual people and their actions are only starting points from which stories can grow, develop, and evolve. For example, Brandenburg's actions were used as a vehicle to move from an individual to collectives of vaccine skeptics in medical professions. Because characters and their actions go hand-in-hand, the action wedge is also implicated by how narrowly we identify our characters and vice versa.

THE ACTION WEDGE

The action wedge refers to the plot in terms of how characters behave and interact related to the flow of events in the narrative. H. Porter Abbott makes sense of action through grammatical structure to propose that "as soon as we follow a subject with a verb, there is a good chance we are engaged in narrative."[19] Charting a story's action wedge reflects to what extent the story's characters are acting logically in accordance with the story (i.e., narrative probability) and whether there is rising and falling action. The action wedge is related to the character wedge because it evaluates whether the events of a story are a result of what communication scholars Michael Dahlstrom and Shirley Ho call "the actions of specific characters" and have been purposefully performed.[20] Consequently, the action wedge evaluates cause and effect and the "necessary relations" between parts of a narrative, including the behaviors of the story's characters.[21]

The action wedge is, in part, mapped by asking, "How likely is it that the story unfolded in a particular way?" Phrased differently, a story's narrative probability can be evaluated based on the consistency of actions with character expectations, the setting, and previous events and forthcoming events in the story. A low probability story would have characters behave in ways that do not align with audience expectations or do not clearly move from event to event. For example, Japanese Kabuki has a rich tradition of color symbolism, by which characters' features, expressions, age, and role in the performance are reflected in their masks and costumes. Those dressed in white represent "unblemished purity," the "highest rank," and "gentility" while "purple and gray lines" painted on the face or on a mask "suggest malevolence or guile."[22] Having a character dressed in white yet acting malevolently and taking on the role of the story's villain may disrupt narrative probability by breaking audience expectations of the mask's meanings.

If the action of a story is logical and expected, and contains concrete behaviors and actions, it would be mapped at the micro-ring. Alternatively, a story that is confusing and has low narrative probability might be mapped on the meso-ring, but a story with no discernable events, actions, or behaviors would be mapped on the macro-ring. It would be an overgeneralization to only consider stories that have clear actions to be "good" stories. While Walter Fisher argued that stories with high probability are more likely to be adopted and accepted, micro-ring actions, much like micro-ring characters, should be considered starting points to ground science storytelling practices. We can modify science's stories to involve a variety of different actions from the concrete and specific to more diffuse actions, changes, and influences.

Embracing meso-ring-level and macro-ring-level mapping on the action wedge recognizes how many stories have iterative effects that amass over time in ways that may not have a discrete, acute occurrence. A narrative with discrete actions may have difficulty capturing the persistent effects of slow violence that "occur gradually and out of sight" and are "dispersed across time and space."[23] For example, deforestation may not be an immediate, high-urgency climate event but can lead to the changing of ecosystems, threatening Indigenous land, and turning oxygen-producing landscapes into capitalist carbon generators that further contribute to climate injustices.[24]

A story with concrete actions may also potentially be limiting in the actions it invites audiences to recreate. In other words, audiences may feel disempowered by narratives that narrow the definition of effective or meaningful actions and leave their capabilities outside of the story's scope. A story that has action mapped on the meso-ring, the macro-ring, or on multiple rings would be able to capture an audience's attention with specific events that occurred while also being able to capture a more accurate sense of the unfolding of stories that do not have an ending or are not discrete.

The following sections outline three features common across institutional scientific stories that are potential barriers to the character and action wedges being mapped on the micro-ring. The three themes are (1) processes and collectives, (2) blameworthy and misinformed counteragents, and (3) fatalism and vulnerability. The first theme addresses how science's stories tend to lack micro-ring characters by focusing on natural processes and collectives or groups as characters. The second theme examines how characters may be present but are ascribed features of counteragents that work against mainstream science, such as participating in blameworthy actions or being misinformed about scientific topics. The third theme analyzes how scientific stories remove agency from audiences who may otherwise be empowered characters in the narrative. Although the character and action wedges do not have to be hubs for any given story, outlining the ways scientific stories tend to use these features illuminates narrative opportunities for audience engagement, especially in the face of rival stories that do use micro-ring characters and actions.

PROCESSES AND COLLECTIVES

Mapping the character wedge on the hub spiral means that a story has rich, specific characters who are similar to the target audience. Science's stories, however, in communicating plain and objective facts, may purposefully remove characters, which rhetorician Leah Ceccarelli attributes to "professional norms that encourage [scientists] to focus on the object of their research."[25] Scientific discourse often "emphasiz[es] results rather than the person completing the action."[26] To say that actions occur but with no person performing them is to remove the human part of the story, which may be less appealing to audiences and may foster a lack of responsibility on the part of scientists for their participation in science.[27]

Sociologist Robert Merton famously described the four norms of science as communism, universalism, disinterestedness, and organized skepticism.[28] Taking the third into consideration for the role of scientific characters, disinterestedness discusses the individual whims of scientists. Specifically, the norm of disinterestedness relies on the concerted effort of science as an institution to police the behaviors of individual scientists through the "accountability of scientists to their compeers."[29] In other words, disinterestedness stems from the actions of the collective scientific enterprise, thereby effectively transcending and erasing individual "preoccup[ation] with financial gain, prestige, or career over the pursuit of knowledge for its own sake."[30] Therefore, norms of scientific discourse may purposefully frame actions as stemming not from the interests and motivations of individuals but of the institution as a discursive way to perform disinterestedness.

Scientific stories frequently lack acting characters who make purposeful decisions. Scientific storytellers tend to center collective groups and behaviors and discuss "public" action or "global" efforts writ large. In some retellings, large abstract groups such as entire scientific disciplines may be the story's main characters. In other versions, scientific stories lack any mention of a character, either through using passive voice or describing natural processes. Rival stories, however, often anchor their narratives in character and action hubs, providing specific examples for audiences to follow and relate to.

Climate change's stories lack characters when environmental processes occur without the purposeful action of a group, person, or acting agent. For example, an Intergovernmental Panel on Climate Change report noted that "adaptation and mitigation are complementary strategies for reducing and managing the risks of climate change."[31] Although there are actions present, reducing and managing, there are no characters performing such actions. Chris Manktelow and I proposed a rephrasing of this statement to be "Global governments and leaders can enact both adaptation and mitigation as complementary strategies for reducing and managing the risks of climate change."[32] The revised sentence includes explicit characters who are performing, albeit vague, actions with a clear goal of addressing climate change. Even though governments and leaders are relatively abstract, they are present characters that communicate the performance of actions within the climate narrative.

Other statements in climate change's stories describe changes in the environment with no attributable cause or human element. For example, the Nature Conservancy wrote that "the average temperature of the earth's atmosphere is rising. As the temperature rises, various impacts are changing aspects of our climate."[33] A "Climate Change Q and A" published by the Royal Society notes that "Earth's average surface air temperature has increased by about 1°C (1.8°F) since 1900."[34] While both publications go on to talk about humanity's role in this warming, these opening statements, the likely first read by audiences, remove specific characters involved in Earth's warming from the story.

To the extent that scientific stories of climate change involve characters, they are vague collectives rather than recognizable individuals. Science as a discipline is sometimes an acting character or is the implied character performing research. The American Association for the Advancement of Science noted, "Observations throughout the world make it clear that climate change is occurring, and rigorous scientific research demonstrates that the greenhouse gases emitted by human activities are the primary driver."[35] On NASA's website about the scientific consensus on climate change, there are many examples of meso-ring characters, including "research has found," "multiple lines of evidence indicate," and "scientists have known for some time."[36] Speaking collectively about climate change emphasizes the vast amount of support for the scientific consensus, which can be an effective rhetorical choice in telling climate change's stories. However, this choice downplays the ability for character or action to be a hub wedge because there are no individuals observing, researching, or acting based on the collected evidence.

In telling the stories of how humans came to be, evolution's narrative distinctly lacks *human* characters. The stories science tells about evolution involve billions upon billions of nonhuman agents living, reproducing, and dying with little change from generation to generation. Instead of using terms related to characters and their actions, many scientific explanations of evolution used agency-focused language that emphasized the means by which evolution occurred. For example, geological scientist Donald Prothero's *Evolution: What the Fossils Say and Why It Matters* described natural selection as a "mechanism" and "engine" that drove evolution.[37] Instead of emphasizing minds, relationships, and purposeful choices, these descriptions focus on mechanistic "substances, processes, and materials," which Burke argues is customary of scientific discourse.[38]

Stories of evolution often downplay purposeful actions by arguing "natural selection has no intentions or senses" and "natural selection has no foresight or intentions."[39] Some scientific texts use passive voice to describe the process of natural selection. For example, anthropologist Cameron Smith's *The Fact of Evolution* notes, "Variations that are beneficial tend to be passed on (*they are selected for*), and variations that are not beneficial tend not to be passed on (*they are selected against*)."[40] In the stories of evolution, there is no acting agent that does the selecting, hidden by the swallowed clause in the passive voice.

Without a character (human, god, or otherwise), nature itself becomes the "selector" in the stories of evolution. For example, Charles Darwin purposefully used terms in *On the Origin of Species* that personified nature to imply that evolution contained elements of "foresight, intelligent planning, and craftsmanly skill."[41] In a section of *On the Origin of Species,* Darwin writes:

> Man selects only for his own good; Nature only for that of the being which she tends. . . . It may be said that natural selection is daily and hourly scrutinizing, throughout the world, every variation, even the slightest; rejecting that which is bad, preserving and adding up all that is good; silently and insensibly working.[42]

It is thus our natural environment that scrutinizes options and makes decisions regarding the evolutionary process. Understanding Evolution uses similar language that gives decision-making power to nature in noting natural selection "simply selects among individuals in a population" and Prothero noted that unfavorable traits are "weeded out by natural selection."[43] These descriptions treat natural selection "as an all-powerful, all-pervasive force," wherein there is little to no room for human characters similar to potential audiences.[44]

Like climate change, the stories of evolution often contain collective characters. The National Academy of Sciences' description of evolution reads "hundreds of laboratory experiments have shown" how chemical compounds can become the "building blocks of life."[45] The National Academy of Sciences additionally reported that "the best estimates of Earth's age are obtained by calculating the time required for development of the observed lead isotopes in Earth's oldest lead ores."[46] These quotations focus on the quantity of experiments conducted but do not name who or what is performing the experiments or calculations. Individual scientists are mostly left out of these descriptions, while scientific disciplines are given voice and agency. A report titled *Science, Evolution, and Creationism* notes that "astrophysics and geology have demonstrated that the Earth is old enough for biological evolution to have resulted in the species seen today" and that "physics and chemistry have led to dating methods that have established the timing of key evolutionary events."[47] Scientists are replaced by the fields that they work in, subsuming the work of individuals into science as a collective term.

In the stories of vaccination, the "action" could be conceptualized as the vaccination process itself or simply "vaccinating" someone. Who is performing this act? In many descriptions of vaccination from scientific and medical texts, the passive voice is used, meaning that the agent is hidden in favor of phrases such as when the American Academy of Family Physicians notes that a vaccine "is given" or "your child usually receives their first vaccine soon after they are born."[48] These phrases do not emphasize parents or individuals as making a vaccine decision or even the doctor that may be administering the vaccination. The character is hidden and the process of vaccination, as if a natural occurrence or foregone conclusion, occurs without human intervention.

When there are characters mentioned in vaccination's stories, the agent is oftentimes collective in nature. For example, an information sheet about the influenza vaccine notes that the "CDC recommends everyone 6 months of age and older *get vaccinated*."[49] Not only does this sentence deflect from doctors who would be performing the vaccination, it also hides parents from view as a decision-maker for a six-month-old. The character is not a specific person or parent choosing to vaccinate, but a group, the Centers for Disease Control, which provides recommendations. The use of passive voice is common to instructional communication, where the focus is information transfer and education on new knowledge.[50]

Government bodies may use instructional communication to address large portions of the population to give advice on a variety of health and safety issues. While this phrasing may be common in the transfer of information, it decenters specific characters and actions, which compromises narrativity. In a study that compared communicating health information on cervical cancer through narrative and nonnarrative formats, communication scholars Sheila Murphy and colleagues found that narratives were more effective than nonnarratives in increasing information retention.[51] Additionally, participants who identified as the same race as the narrative's main characters resonated most with the story and experienced the strongest emotional response compared to other participants.[52] Simply relying on the direct transfer of information without characters and other narrative features may leave scientific storytellers at a disadvantage to those who more directly incorporate narrative elements such as characters.

A page on the Healthy Children's website notes, "Both the AAP [American Academy of Pediatrics] and the Centers for Disease Control and Prevention (CDC) want as many children as possible to get a flu vaccine each and every year," characterizing both organizations as acting and feeling agents with a desire for widespread vaccination.[53] A World Health Organization's webpage on COVID-19 notes, "WHO's primary focus now is to support countries to turn vaccines into vaccinations as fast as possible." This quotation describes the collective character of "countries" as taking unspecified actions to vaccinate their populations.[54] The Centers for Disease Control and World Health Organization, of course, are not individuals themselves, but are conglomerates made up of individual experts acting in concert. Expressing collectives in this way can confer a sense of authority and credibility onto the institution by subsuming individuality but can also hide individual, specific characters from the narrative.

In the case of COVID-19, the Centers for Disease Control website refers to itself as "continuously monitor[ing] vaccine effectiveness to understand how COVID-19 vaccines protect people in real-world conditions."[55] However, the Centers for Disease Control coronavirus website removes itself from the position of acting agent in noting that "research shows that the COVID-19 vaccines used in the United States continue to protect against severe diseases, hospitalization, and death from known circulating variants."[56] In a likely attempt to keep the websites "neutral" and fact oriented, the Centers for Disease Control has scrubbed mention of individual people who might serve as characters. This framing also positions audience members as passive recipients of vaccines by a nebulous group, as will be discussed in the "fatalism and vulnerability" section.

A straightforward fix for the lack of acting characters would simply be to insert them into scientific stories. In some circumstances, this may be a relatively easy task to locate a particular scientist or affected community to emphasize the specific unfolding of events by individuals. However, rhetorical decisions to highlight particular people and events are not without narrative risks themselves, including inaccuracies and gatekeeping. No scientific controversy is a matter of only individuals. Science is a communal process and scientific topics affect everyone all over the world. Narrowing characters and their actions to specific instances can thereby undermine the scientific process by overly restricting our focus to heads of scientific organizations or spokespeople, who are coded as "scientists." The work of larger groups and the more diverse members therein may be overshadowed.

Inserting humans into the stories of evolution, for example, would likely be inaccurate. Humans are not present to serve as characters for much of evolution's history. One option would be to use meso-ring characters, such as anthropomorphized animals, as examples of evolutionary change over time. While nonhuman animals do not perform the action of evolving, they do have motivations related to their survival instincts that guide their actions, which could be discussed in a narrative.[57] Biologist Sara ElShafie characterized evolution as having nonhuman animal "protagonists" that may drive the plot of the story through their decisions.[58] ElShafie notes that describing animals' behaviors should not be done in a scientifically "inappropriate sense," but as a way "to allow an audience member to find meaningful parallels between [their] own experience and that of the nonhuman character."[59] In other words, a stories of evolution with nonhuman animals could clearly signal that there are differences in the actual functioning of evolution while still using nonhuman animal characters to capture an audience's attention. Another option would be to use a micro-ring character of a scientist doing research about evolution or a layperson learning about evolution, such as a student in a classroom. While the content could be similar, the framing of the story and the characters within would be quite different from an instructional piece of information that lacked explicit characters and actions.

While Western inquiry may call attributing human characteristics to more-than-humans a "fallacy,"[60] some cultures view such personification as natural and accurate. Indigenous narratives regularly personify nonhuman animals and parts of nature, such as the sky, stars, and Earth, and consider the natural environment to communicate with people, if only we would listen.[61] Folklorist María Inés Palleiro argued that such narratives exist in a "gap between reality and fiction" and are "rooted in real experiences [and represent] discursive attempts to explain" encounters and interactions.[62] Specifically, "animal tales" such as Aesop's fables or southwestern stories of the "Spider Woman" represent various interactions of reality and fiction that function to explain and persuade.[63] Using a meso-ring for characters in a story of evolution can make room for nonhuman characters to act and be a part of the narrative. Although not as specific as individual people, meso-ring characters can be more accurate to the science and appeal to different conceptualizations of human-nature interactions.

Having characters similar to audiences, namely, individual people, does not guarantee an effective narrative. Another dynamic within the character and action wedges emerges in the resonance between audiences and the characters and actions portrayed in the narrative.

BLAMEWORTHY AND MISINFORMED COUNTERAGENTS

In some circumstances, a story's failure to resonate, or low narrative fidelity, can be attributed to audiences feeling misalignment between the characters represented and themselves. A study by communication scholars P. Sol Hart and Erik C. Nisbet found that people were more likely to support climate mitigation policies when they identified with victims discussed in a climate narrative, whereas a lack of identification and political partisanship decreased support.[64] Similarly, a study by communication scholars Megan R. Dillow and Keith Weber found that demographic similarities, such as living in the same state, increased people's likelihood of signing up to be an organ donor, which the researchers attribute to participants being able to see themselves in future need.[65] In a series of studies I performed with colleagues at the University of Nevada, Las Vegas, and the University of Wisconsin, Madison, we found that a short introductory exercise where group members found two things in common with each another led to increased success in a climate policy task, more positive emotions, and shifting attitudes toward thinking that climate change is real.[66] In short, identification between oneself and others can be a powerful rhetorical tool to bridge gaps, collaborate, and encourage behavior change.

To leverage the tool of identification for scientific stories, one strategy may be to locate specific individuals as characters. However, scientific stories with micro-ring characters tend to position audiences as those who have intentionally or unintentionally made choices that contribute to the controversy, what Burke would call being "counter-agents." For Burke, agents are the characters in a drama who perform actions. Agents can have "friends (co-agents) or enemies (counter-agents)" who influence the performing of the actions by agents.[67] When they portray people, such as ourselves, as bad or as working counter to the main characters of a narrative, micro-ring or meso-ring stories that emphasize negative actions, whether intentionally or unintentionally, can fail to gain adherence by audiences.

In the stories of climate change, humans create new technologies and build industrialized cities, and, in the process, destroy our earthly home. People in climate change's narratives are framed as responsible for many of the predicted negative consequences of climate catastrophe, from extreme weather events to droughts to heat waves to changing migration habits. For people who do not see themselves as polluters or as having agency over the environment, this framing as a blameworthy counteragent may not resonate with them. Sometimes directly and sometimes indirectly, climate change's stories place blame on our industrial activities and environmentally destructive behaviors that require collective attention and concerted effort to fix.

In many climate stories, the audience must come to terms with their counteragent status by making sacrifices to correct humanity's past and ongoing wrongs. The American Physical Society's "Statement on Earth's Changing Climate" noted "multiple lines of evidence indicate that human influences have had an increasingly dominant effect on global climate warming" and that these influences "are growing."[68] At another point in its statement, the American Physical Society described "the climate impact of human activities." These statements link human behaviors and activities as the cause of various effects and impacts on the environment. The American Meteorological Society noted that "research has found a human influence on the climate" and that "climate can be affected by human-induced changes in atmospheric composition."[69] Although this particular usage of "human" positions it grammatically as a modifier instead of an agent, the implication is clear that climate changes are caused by people.

The National Academies were more direct in their characterization, noting "scientists have known for some time, from multiple lines of evidence, that humans are changing Earth's climate, primarily through greenhouse gas emissions." In NASA's publication "The Causes of Climate Change," the relationship between human behaviors and climate impact is clearly expressed: "Humans have increased atmospheric CO_2 [carbon dioxide] concentration by more than a third since the Industrial Revolution began. This is the most important long-lived 'forcing' of climate change."[70] NASA's phrasing evokes human responsibility for environmental changes and quantifies the impact. Furthermore, the use of "forcing" connotes an unnatural, aggressive interaction between humans and the Earth.

The Environmental Protection Agency's briefing on climate change describes human activities as "the dominant cause" of warming and as having "contributed substantiality to climate change."[71] The Environmental Protection Agency further notes, "The more we emit, the larger future changes will be," emphasizing that even slight increases to carbon-emitting behaviors will still have impacts.[72] The Environmental Protection Agency notes in a different online publication that "greenhouse gases from human activities are the most significant driver of observed climate change."[73] Descriptors such as dominant, substantial, and significant add more heft to the causal link between human behaviors and negative impacts. The Environmental Protection Agency also used analogies to further visualize humanity's effects. Comparing a bathtub faucet to the production of CO_2 emissions, the Environmental Protection Agency noted that "human activities have turned up the flow from the CO_2 faucet," putting the bathtub at risk of overflowing.[74] Attributing climate change to human behaviors is scientifically accurate, and we want to retain accuracy, but it is also important to recognize that these decisions also tell a story that emphasizes humanity's blameworthiness over environmental disruption.

Many climate texts referred only vaguely to the specific blameworthy activities we have participated in, but others more directly named them. For example, the American Meteorological Society described how the greenhouse gas methane is produced "from human activities such as agriculture, landfills, and fossil fuel extraction processes, [which are] the human activities *responsible* for the majority of emissions today."[75] NASA specifically calls out industrialization as a cause of climate change: "The industrial activities that our modern civilization depends upon have raised atmospheric carbon dioxide levels from 280 parts per million to 412 parts per million in the last 150 years."[76] Similarly, one scientific text from the National Academy of Sciences noted that human activities, "especially the burning of fossil fuels since the start of the Industrial Revolution," bear the largest responsibility for climate change.[77] The Environmental Protection Agency also notes that the uptick in carbon emissions is due to "burning fossil fuels," which people use "to generate electricity, heat and cool buildings, and power vehicles."[78] These statements directly link modern comforts to the degradation of the environment.

By having their activities listed as blameworthy, humans become the counteragents of climate change's stories who must be rehabilitated and change their ways to prevent future warming. Scientific texts about climate change mostly focus on the causes and contributors to climate change, but some did discuss the necessary sacrifices that must be made and the importance of doing so. The Environmental Protection Agency noted that climate change will "require adaptation on larger and faster scales than in the past," which they acknowledge will present "challenges to human well-being and the economy."[79] Even though the Environmental Protection Agency report lists the negative consequences of unmitigated climate change, it also notes the "risks" inherent in adaptation and mitigation to the standard practice of everyday life. In its fifth report, the Intergovernmental Panel on Climate Change argued that "lifestyle and behavioural [*sic*] changes could reduce energy demand" as well as "changes in consumption patterns."[80] Simply put, many of the scientific texts call for "reducing the amount of greenhouse gas emitted by human activity" as the "only way to avoid much of the projected warming."[81]

Positioning human behaviors as causes and contributors places a burden of responsibility to participate in mortification, or self-sacrifice, which people can be hesitant to do if it means giving up comfort and convenience. In the case of climate change, mortification "implicates humanity as a whole" because we have all contributed to burning fossil fuels in our daily activities.[82] For example, the fifth report from the Intergovernmental Panel on Climate Change outlined how individualism is counterproductive to climate progress: "Effective mitigation will not be achieved if individual agents advance their own interests independently."[83] The U.S. Global Change Research Program noted in its climate assessment that a "substantial and sustained global mitigation" program is needed "to avoid the most severe consequences [of climate change]."[84]

Climate skeptical groups and anti-environmental advocates are quick to note that climate change discourse places blame on human behaviors. For example, the Cornwall Alliance, a Christian fundamentalist group, argues that environmentalism "rests on and promotes a view of human beings as threats to Earth's flourishing."[85] Journalist Brendan O'Connor called the Cornwall Alliance a religious "enforcer" that keeps the Christian, and specifically Evangelical, community in line in order to stop environmental attitudes from "gain[ing] a following" within the faith.[86] Another influential group, the Heartland Institute, a conservative think tank, argues that environmentalism "blames humans for global warming and climate change,"[87] instead of treating warming as part of the Earth's natural warming and cooling cycles.

If we accept climate change's stories, we also accept the responsibility and guilt that accompany environmental degradation. Philosopher Bruno Latour argued that people "have difficulty submitting" to climate change's narratives because they are "supposed to feel responsible" for the Earth's current condition.[88] People do not want to think of themselves as malevolent and also do not want to compromise their well-being and comfort. Geographer Anthony Leiserowitz found that 90 percent of survey respondents agreed with the statement that the United States should reduce its carbon emissions, but a majority of respondents opposed specific policy changes that were "direct pocketbook issues," such as levying a business tax (31 percent support) or a gas tax (17 percent support) in order for reductions to occur.[89] Even those who are concerned about climate change find it hard to make personal sacrifices. In emphasizing the causes of climate change and focusing less on potential solutions, climate change stories highlight our status as counteragents to environmental flourishing instead of our potential heroic redemption.

Evolution's stories have previously been discussed as lacking human agents and purposeful actions. Additionally, some retellings of evolution emphasize the evolutionary process as a competitive fight for survival. Poet Alfred Tennyson (1850/2013) famously wrote that "nature is red in tooth and claw," meaning that it is through blood and death, and not discourse or politics, that life has evolved.[90] Answers in Genesis, a creationist apologetics nonprofit, publishes frequently on evolution's violent past as a reason to reject evolutionary science. Answers in Genesis writes that evolution "reduces human beings" to "mere players in the game of survival of the fittest." Instead of humans being made in God's image, Answers in Genesis argues that evolutionary science promotes "continual death and struggle" that fosters mass shooter mentalities and "senseless violence."[91] In another article, Answers in Genesis CEO Ken Ham argues that

> the public school system has indoctrinated generations of young people into believing there's no God, we're just highly evolved animals who evolved over millions of years, and there're no moral absolutes. And in this worldview where we decide right and wrong for ourselves, then violence is okay if I think it's okay.[92]

Some people do not want to see themselves as animals and thereby may not find evolution's stories to be faithful to their understanding of the world. To say that humans evolved from nonhuman animals implies that we are consubstantial with a violent, cruel past. For some, it is much more appealing to think of humans as separate from this violent nature and more than the random by-product of billions of animals struggling to survive.

The stories of vaccination and COVID-19 position at least some portions of the population as counteragents, namely those who refuse to get vaccinated, delay vaccination, or refuse COVID-19 preventative measures. Although it may be tempting to lay the blame of continuously rising deaths and illnesses at the feet of the unvaccinated and anti-vaccine community, such positioning does little to engage their decision-making and alienates these communities from changing their minds on the topic.[93] The Centers for Disease Control uses blameworthy language in addressing the role of parents as potential counteragents. The site says, "If you don't vaccinate, know your responsibilities," including "it's your responsibility to inform your child's school, childcare facility, and other caregivers about your child's vaccination status" because "they need to consider the possibility that your child may have a vaccine-preventable disease."[94] The web page continues, "A decision not to immunize your child also involves risk and could put your child and others who come into contact with him or her at risk of contracting a potentially deadly disease."[95] A website from the American Family Physician echoed a similar counteragent framing in stating, "The choice of some parents not to immunize increases the infection risk for all children, including those who are immunized," situating parents who do not vaccinate as counteragents to the act of vaccination and as threats to public health for all children, including their own.[96]

Many anti-vaccination messages reference the right to choose and how mandates restrict parental choices. Many scientific organizations, however, praise these restrictions and subsequently, some may interpret, the disempowerment of individuals over certain health decisions. For example, the American Academy of Pediatrics does acknowledge that there are "appropriately implemented medical exemptions" to vaccination, but the group also noted that it "opposes nonmedical exemptions to specific immunization requirements because it leads to lower immunization rates and increases the risk of spreading dangerous vaccine-preventable diseases."[97] The use of "appropriate" indicates that there are some medical exemptions that are inappropriate and thus should be rejected as reasons not to vaccinate. This also suggests that any exemption that is nonmedical, such as personal or religious beliefs, should be rejected.

Similar messaging was expressed by the American Academy of Family Physicians, which wrote that the group "does not support immunization exemption policies except in cases of allergic or medical contraindication."[98] Medical doctors Jeanne Spencer, Ruth Trondsen Pawlowski, and Stephanie Thomas note, "In the absence of a direct threat from disease, it is clear that some patients and parents will not participate in vaccination programs unless absolute safety can be assured."[99] This statement implies parents are misinformed about vaccines because they do not see the direct need for them in their communities. In context, the American Academy of Family Physicians considers wanting absolute safety unreasonable and casts vaccine hesitant parents as misinformed and thus making poor health decisions.

Although vaccine hesitant parents are not often characterized as intentional wrongdoers, they do act as misinformed counteragents who work against the goal of scientific groups to increase vaccination. For example, the website Vaccinate Your Family argued

> you may have heard that vaccines contain all types of crazy ingredients that sound as though they don't belong in a medical product. The truth is that a very small group of very vocal, but misinformed, individuals have made false accusations regarding the safety of vaccines and their ingredients. In most instances these allegations are just incorrect. In other cases, the claims are misinformed or taken out of context.[100]

This quotation frames parents who follow these voices as being "misinformed" and believing in false accusations. Despite attempts to characterize vaccine hesitant parents as simply misguided, some parents may feel like "they are becoming pariahs in their communities" due to their vaccination decisions.[101] Communication scholar Shari Hoppin attributes this hesitancy in part to the compelling and convincing narratives of vaccine hesitancy that fulfill parents' "desire for agency" over their children's health and well-being.[102]

Even more strident messaging emerges in COVID-19 discourse. One scientist, a drug expert at Australian National University, called advocating for alternative medicine such as hydroxychloroquine "insane."[103] An epidemiologist at Columbia University referred to Trump's handling of the pandemic as "not just ineptitude, it's sabotage."[104] This type of discourse was rare from scientists themselves, however, meaning that science storytellers from technical spaces did not tend to portray COVID-19 skeptics as villains or counteragents, at least not in inflammatory language. Shaming of COVID-19 skeptics or those seeking alternative treatments circulated more prominently in public discourse. Some supporters of vaccination and mask policies shame and call for repercussions for others who do not follow guidelines.[105] While these storytellers are not scientists, they are science communicators who are part of the larger narrative ecology of circulating stories in the COVID-19 controversy. In other words, not only formally trained scientists tell scientific stories; lay people do so as well in ways that can contribute to people's understanding of science and what the "science" says.

Politicians are often not themselves trained scientists but can become de facto spokespeople for scientific messages in their role as bridges between the technical and public sphere. For example, early in the pandemic, North Carolina Governor Roy Cooper (D) was quoted as saying that people's "refusal to wear a mask is selfish" and goes against "basic decency and common sense," further noting that people who do not want to wear masks should not go into public places like grocery stores.[106] The frustration that many people feel toward COVID-19 skeptics or folks who are unwilling to mask or vaccinate is certainly understandable. Voicing those frustrations, however, may inadvertently (or advertently) highlight divisions within the population. Sometimes, these divisions turn violent, such as the case of a teacher in Amador County in Northern California who was attacked by an anti-masking parent, or when a bus driver in New York City was punched in the face by an unmasked man who was asked to put on a mask before boarding.[107]

Stories of conflict can be categorized as "melodramatic" in that they highlight wars between good and evil, agents and counteragents, heroes and villains. This discourse can be highly captivating and mobilize supporters to fight urgently for good, especially in environmental contexts.[108] However, war frames do have narrative drawbacks. Even though particular people and groups are named, a micro-ring mapping on the character wedge that relies on creating in-groups and out-groups (e.g., climate skeptics, climate believers; creationists, intelligent designers, evolutionists; vaccinated and unvaccinated; masked and unmasked) can shut down productive communication and collaboration. When we turn our war frames to each other, we reinforce xenophobia and racial differences between people.[109] For example, there was an increase in anti-Asian incidents and hate speech during the COVID-19 pandemic, which many attribute to terms such as the "China virus" and "Kung flu" that circulated in reference to Wuhan, China, as the starting point for the COVID-19 outbreak.[110] Additionally, these sweeping labels lump complex groups such as "climate skeptics" and "the unmasked" into the same category without regard for situational, cultural, economic, or social differences.[111]

Identifying villains in a narrative is a rhetorical choice that comes with potential benefits of engaged audiences and urgent responses in the short term, but can also easily reinforce boundaries and stall collaboration in the long term. If our goal of telling scientific stories is to promote change and action, then we do not want to narrow audiences into particular roles that may cause them to reject the narrative or feel disempowered to make changes.

FATALISM AND VULNERABILITY

When people are incapable of acting, they lack agency, feel powerless, and may be fatalistic toward the ability for any change to be enacted. Burke's definition of agency is related to the "instruments" used to perform an act.[112] From the more obvious tools of a hammer, nail file, or piece of technology, Burke also included "intelligence" and "hands."[113] Restricting someone's agency, therefore, is similar to removing their conceptualization of self in addition to limiting their movement, expression, and thoughts.

Philosopher and gender theorist Judith Butler argues that identity is not static or fixed but is constituted by "acts and gestures" that are interpreted within societal norms.[114] She further argues that the identity of a "'doer' is variably constructed in and through the deed" and does not exist *a priori* to the performance of acts.[115] Similar to Butler, cultural theorist Stuart Hall sees identity as inextricable from societal norms that constrain identity performances. Hall's focus is on both "discourses and practices" that "articulate" an identity.[116] Without our having the ability to speak, act, contribute, or participate, our senses of identities may be fractured, fragmented, excluded, and, ultimately, constructed by others instead of ourselves.

In the case of climate change, there has been an increase in reports of "climate anxiety" or "eco-anxiety," where people are concerned about present and future consequences of climate change and skeptical of our collective ability to forestall catastrophe. Social scientists Julia Sangervo, Kirsti M. Jylhä, and Panu Pihkala defined climate anxiety as "feelings of tension, worried thoughts, and physiological changes" related to the climate crisis.[117] Climate anxiety can reduce one's sense of purpose, activate avoidance behavior, and foster a perceived inability to act.[118]

A focus on individuals' environmental actions is not accidental; such feelings can be traced, in part, to deliberate marketing campaigns by heavy polluters to shift blame for carbon emissions onto individuals. In 2004, British Petroleum unveiled the carbon footprint calculator for people to measure how much they are personally contributing to the climate crisis through daily consumption behaviors, commuting, and other travel.[119] In what seemed to be an eco-friendly campaign to encourage people to make thoughtful decisions about carbon consumption, the calculator also had the intended consequence of shifting the focus away from companies such as British Petroleum, which are overwhelmingly responsible for carbon emissions to the tune of nearly 90 percent.[120] Journalist Amy Westervelt called this strategy the "weaponiz[ing of] American individualism," which thereby deflected attention from structures that enable individual consumption.[121]

Part of the counter-messaging to the individualizing messages of the carbon calculator forwarded the idea that individual decisions are *not* meaningful dents in overall carbon emissions. Overcorrecting for the carbon calculator may have the inadvertent effect of increasing climate anxiety and reinforcing a lack of agency for individuals. In an article published on *The Conversation,* engineer Marcelle McManus argued, "It may no longer be in anyone's personal capacity to make changes great enough to reverse the damage already done."[122] Even worse than being trivial, Jay Michaelson, columnist for the *Daily Beast,* argued that calling for individual behaviors to stop the climate crisis is "counterproductive," "hurting," and little more than "distraction."[123] Hearing refrains of individual actions being meaningless, audiences may feel disempowered and vulnerable to a changing climate and a system made up of larger, more powerful characters.

A similar sense of meaninglessness emerges in the stories of evolution. Evolution is not a purposeful act dictated by the decision-making of creatures, but a random "concatenation of motions" influenced by the environment over time.[124] One scientific summary of evolution noted that "chance and randomness do factor into evolution and the history of life in many different ways."[125] Life is not purposeful, designed, or reliable; one author described life as "in essence being a giant agglomeration of atoms," which is "inherently unpredictable."[126]

In evolutionary biologist Richard Dawkins's *The Blind Watchmaker,* he describes natural selection as a "blind, unconscious, and automatic process" with "no purpose in mind" that "does not plan for the future."[127] Comparing the stories of creationism with evolution, Dawkins notes that the story of divine intervention "involves design and benevolence" while evolution contains neither and "only a purposeless process of attainment or failure."[128] In *The Fact of Evolution,* Smith noted,

> Evolution is the ball rolling down the slope; it is a consequence, not a plan or even a "thing." It just happens. Because of the replicating, variable nature of lifeforms, and the dynamic, changing nature of the universe, evolution has to happen.[129]

Evolution is not a purposeful act dictated by decision-making, but a random process of genetic combination that produces slight changes over time. Evolutionary scientist Keith Fox argued that life only looks designed because "we are looking at the finished product," when in fact "meaning and purpose" are not present in evolution.[130]

If evolution and natural selection have no purpose, then life itself and humans as part of life may be nothing more than "the chance results of history" and "historical accident[s]."[131] Humans do not emerge until the very end of the story, a product and outcome of many acts beyond their control. The Smithsonian Institution's Human Origins Program noted, "Modern humans are the product of evolutionary processes that go back more than 3.5 billion years, to the beginning of life on Earth. We became human gradually."[132] A NASA publication summarizing the results of an experiment argued that "humans might not be walking on Earth today if not for the ancient fusing of two microscopic, single-celled organisms."[133] These descriptions communicate how unlikely human life is. Chance is framed as the determiner of life's fate, displacing people as agents and removing control over their actions. However, people like to feel that they make decisions, act, and have control, at least somewhat, which can lead to low resonance between evolution's stories and audience expectations.

Stories about vaccination displace parents and guardians from decision-making capacities, thereby removing an audience's ability to see themselves as in control. When stories deflect and displace parents from the narrative, there may be a sense of lost control or that parents are ancillary to the vaccination process. Historian of science Mark Largent argues that the "medicalization" of child-rearing in the United States has led to parents feeling "compelled" to vaccinate their children to meet pediatricians' and childcare facilities' expectations.[134] The American Academy of Family Physicians argued "everyone needs vaccines," and the American Academy of Pediatrics encourages childcare facilities to "get as many staff and children immunized before and during the flu season as possible."[135] These statements paint a picture of nebulous organizations and faceless groups prescribing universal vaccination that do not directly include a role for parental approval or oversight.

When parents are making decisions to vaccinate their children, they are often "weigh[ing] what they perceive as the costs and benefits of vaccines."[136] When vaccination information is given directly and seemingly without room for adjustment, people may feel constrained in their ability to act. Family Doctor argued that "vaccines are generally safe" and that "the protection provided by vaccines far outweigh the very small risk of serious problems." While these statements may seem encouraging of vaccine safety on the surface, they may also undermine parental decision-making in weighing those risks for themselves. If parents do not see that they have a say in the process, they may feel disempowered as agents in their child's health care or feel locked into a particular course of action. Lacking control and the ability to act can undermine parental feelings of security in their role over their families' health and make them question the decision-making of others, such as the Centers for Disease Control, that has supplanted their own.

To cut off parents' agent status is also to curtail their role in the action of the narrative. When a child becomes sick or is diagnosed with particular disorders, such as autism, parents and guardians seek answers to explain their situation. The idea that a variety of uncontrollable, random factors has affected the health of their child may encourage people to see themselves as failing in their primary role as parents, unable to find a reason for or a solution to their child's perceived ailment.[137] Alternatively, a story of a vaccine harming one's child provides a single, understandable act and complete explanation of a diagnosis. Within the vaccination narrative, narrative probability is low, because there is no tidy explanation for a child's diagnosis or a way to "solve it." Many anti-vaccine advocates (falsely) propose that alternative medicines and treatments can solve for diagnoses such as autism, thereby offering families solutions as opposed to a permanent diagnosis.[138] A story that lacks agency for parents may fail to provide the comfort that stories are expected to provide and may not resonate with parental desire to take care of and heal their children.

People may feel that the stories of vaccination deny their identity as parents to make health decisions. Marcella Piper-Terry, the founder of anti-vaccine group VaxTruth describes organizations such as the American Academy of Pediatrics as "bullies," noting that the organization "pushes pediatricians to have their patients vaccinated and the propaganda is extreme."[139] Even if the statements quoted here by the American Academy of Pediatrics do not appear extreme, they may be interpreted or described as such by the vaccine hesitant. Coupled with circulating myths around vaccine side effects, parents sharing stories of perceived vaccine injuries, and benefits of vaccinating being mostly intangible, the stories of vaccination may not ring true to people's perceptions of relative risks and benefits. Fisher noted that "if a story denies a persona's self-conception, it does not matter what it says about the world," it will be rejected.[140] Vaccination's stories thus complicate the identity of parents by removing them from the position of head of their family's health. Consequently, such a story has low probability because it may seem illogical to take that power away from parents and has low fidelity because people may not resonate with being helpless or sidelined in their children's medical decision-making.

One response to removing agency from audiences is to insert them into the narrative as characters who are empowered to act (or are similar to acting characters). Scientific stories can design specific roles for audiences to play to reduce feelings of anxiety, fatalism, and vulnerability. In some stories, it may be hard to conceptualize such a role. How might audiences become active parts of the stories we tell about evolution? Instead of the process of evolution itself, audiences may be imagined in educational roles as contemporary science communicators who, for example, take their children to science museums. To empower audiences to be involved in science, we can be creative and imagine various parts people can play in communicating science, supporting science, and even doing science themselves.

CONCLUSION

The character and action wedges are some of the first features we think about when we identify something as a story. Considering characters and action in tandem with one another can be a powerful tool for science communication by prompting reflection on their role in a scientific topic or specific content being communicated. Without providing specific prescriptions for scientific stories, which will be situation, audience, and purpose dependent, I offer the following questions to brainstorm the character and action wedges for science communicators and storytellers:

- Does the story have characters? Are they nonliving, nonhuman, or human? Are those characters similar to my target audience or could they be adapted to have relatable characteristics?
- Who are the story's heroes and villains, agents and counteragents? What role is the audience invited to occupy in the story?
- Have I made someone or a group of people villains? Are there ways to retell the story to replace or supplement war frames or do the war frames serve a mobilizing or moralizing purpose?

- Does my story have concrete actions? Are there ways to conceptualize actions with specific examples to illustrate the flow of events to audiences?
- Do I want my audience to be empowered to act in some way? Does my story remove agency or a sense of purpose from my audience?
- Does having concrete characters and actions align with my science communication goals and audience expectations?
- Can I incorporate a variety of character rings by discussing individuals, collectives, and abstract characters?
- Can I incorporate a variety of action rings by discussing specific, time-bound actions as well as more abstract and ongoing actions?

These questions are meant to spark reflection and contemplation as opposed to prescribing an ideal map placement on the character and action wedges. Instead of over-relying on scientific norms or uncritically adopting standard patterns of communication, narrative webs invite us to consider each narrative feature as a rhetorical decision to add specificity or deploy the abstract in describing a story's characters and their actions.

As evidenced by the preceding analyses, the stories of climate change, evolution, vaccination, and COVID-19 told by Western storytellers tend to rely on meso-rings and macro-rings, where characters and actions are vague, abstract, collective, or even absent. Alternatively, as discussed in the case studies chapter, rival stories to science often use micro-ring characters and actions to anchor their stories in specific characters who are like target audiences and empower audiences to act. Rival stories position people as heroes who can make meaningful, powerful changes to their environment and circumstances. Differences in the character and action wedges may, in part, explain the captivating nature of rival stories even in the face of empirical evidence amassed by thousands of scientists across the world.

The goal of this book is not to promote mimicry of rival stories in an uncritical turn to individualism and micro-rings. We do not want scientific stories to fall into the trap of focusing narrowly on individuals and thereby ignoring the larger structural issues that create the conditions in which we act. Productive rival stories can help science communicators emphasize community-level impacts related to climate justice or medical disparities by working from meso-rings and macro-rings. Instead of being overly prescriptive, we can learn from stories that have micro-ring character and action wedges as potential options for adapting our storytelling techniques. Exemplar stories and additional considerations for how to attend to the character and action wedges while avoiding overly narrowing them are explored further in the constellation practices chapter.

By considering the role of characters and action, some science storytellers may locate individuals within their stories and give audiences a sense of purpose to encourage both adoption of the story and future action to be taken. Instead of removing individuals altogether, positioning people as villains, or disempowering audiences, science's stories can embrace a variety of different characters as agents who perform concrete actions. Micro-ring character and action stories may invite audience resonance with characters and empower them to be a part of science's stories.

3 Sequence and Scope Wedges

In the future, sea levels have risen to such heights that nearly all land is completely submerged in water. People who survived the flooding live on floating rafts. With no soil to grow food, most life has perished and the little that remains is constantly on the brink of death. In a different future, the heat of the sun has scorched almost everything to dust. Warring clans battle in a desert wasteland for access to water and greenery, with many perishing in the fight for resources. These are possible futures imagined due to climate catastrophe in the movies *Water World* (1995) and *Mad Max: Fury Road* (2015), respectively. Although the stories are quite different in terms of how climate change's consequences will manifest, they are both dystopian narratives that detail how a changing climate made the Earth inhabitable for life.

Water World and *Mad Max: Fury Road* are examples of science fiction (sci-fi) films that involve climate themes, sometimes called "cli-fi" or climate fiction. Cli-fi and other genres such as eco-horror having been growing in popularity as contemporary representations of our climate anxieties.[1] Unlike documentaries that may show the horrors of the present, cli-fi and eco-horror films can show even more imaginative and ominous disasters of our own making in real time on the big screen.

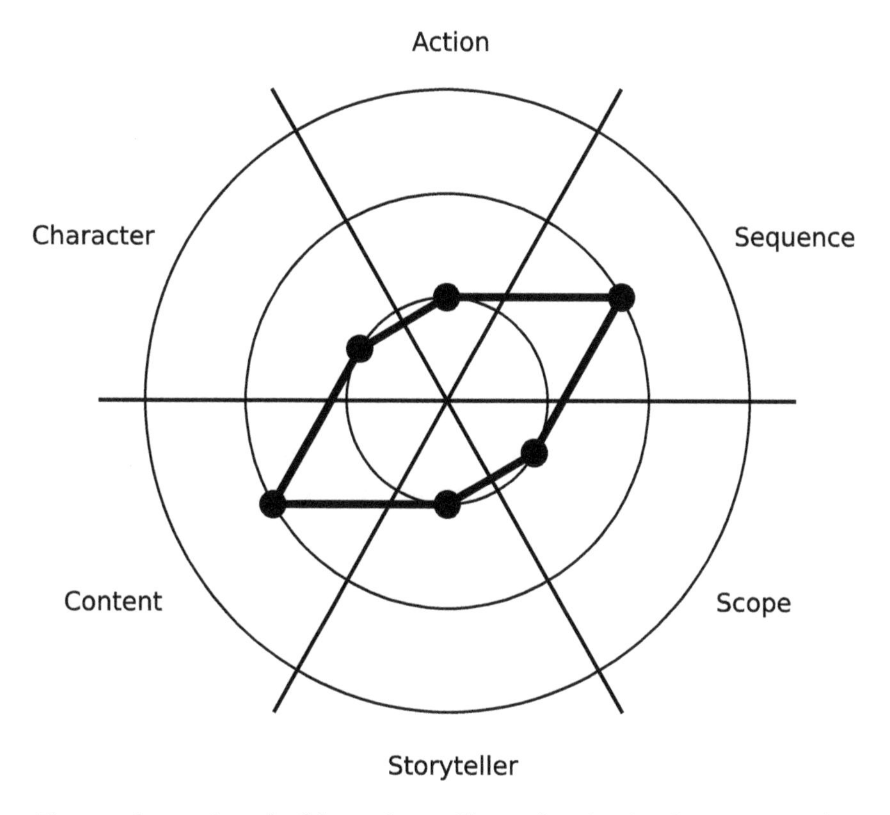

Figure 4. A narrative web of the eco-horror film *mother!* showing the sequence and content wedges mapped on the meso-ring.

Consider, for example, the eco-horror film *mother!*, directed by Darren Aronofsky (2017). The film uses a metaphor of an abusive relationship to illustrate the unsustainable relationship between humans and the Earth, represented by Him, played by Javier Bardem, and mother, played by Jennifer Lawrence, respectively.[2] The scope of an entire planet is thereby reduced to a specific couple and their home, collapsing broad spatial scales into micro-ring ones. The film's narrative is not linear, but cyclical, as the cycle of abuse precedes Lawrence's character and will continue after her. At the end of the film, mother sets the house alight as her only recourse against swarms of trespassers who attack her and the home. Soon thereafter, Him is seen carrying a scorched mother through the ashes of the home. Him assures her, "It won't hurt much longer" to which she responds,

"What hurts me the most is that I wasn't enough." Him replies, "It's not your fault. Nothing is ever enough. I couldn't create if it was. And I have to. That's what I do." This exchange reiterates the inevitability of mother's destruction as it is in Him's nature to destroy. Indicating the narrative's rebeginning, the home reassembles itself and a new unnamed female-presenting character takes the place of mother in Him's bed. *mother!* uses a nontraditional sequence to indicate that the cycle of abuse at the hands of Him and the many invaders is inevitable (see figure 4).

Through our media production and consumption, people try to capture the inherently diffuse impacts of climate change in more overt and tangible ways. Philosopher and ecologist Timothy Morton proposed the term *hyperobjects* to name phenomena distributed across time and space on grand scales that resist localization, such as micro-plastics, capitalism, and the ecological crisis of climate change.[3] While climate change itself is too massive, complex, and multidimensional to be pinpointed, its effects can be felt at tangible, local levels, such as through "climate-driven" disasters.[4] Wildfires burning along the West Coast of the United States are aggravated by warming seasons with less rainfall, a consequence of climate change but not climate change in and of itself. The 2022 floods in Pakistan killed fourteen hundred people and unhoused more than a million others, which United Nations Secretary-General Antonio Guterres called "climate carnage."[5] People may make sense of the fluid, dynamic, and amorphous scope of climate change through the perceptibility, closeness, and likelihood of its effects, on screen or off.

Most living creatures' perceptions are limited to what they are physically capable of sensing, such as through touch or sight. Humans, however, have developed tools to observe at much smaller scales than eyes normally can and grasp information much larger than we can typically comprehend. Uncovering the dimensions of scale is part of the endeavor of science; English scholar Joshua DiCaglio argues "all science moves within and along different scales: whether probing the relationship between quantum motions and stellar phenomena, between DNA molecules and the formation of bodies, or between human actions and ecological effects."[6] It is through our various tools (technological, scientific, and artistic) that humans can escape our limited sensory perceptions and reimagine our world from the infinitesimally small to the broadest of hyperobjects.

The sequence and scope wedges of narrative webs address issues of scale, spatiality, arrangement, and temporality. Together, they form a story's setting—where and when the story is happening, over what duration, and in what order. Sequence attends to the ordering of the events and actions within a narrative from the past and into the future. Scope encompasses temporal and spatial scales from short periods of time to eons of history and from the microscopic to the colossal. Critically reflecting on the setting of science's stories can encourage strategic decisions of how best to represent the chronology and scale of the topic to engage audiences in scientific information that may challenge the limits of our perception and comprehension.

THE SEQUENCE WEDGE

The sequence wedge primarily attends to the ordering of events within a narrative, or what communication scholar Chris Ingraham calls "temporally ordered narrative clauses."[7] Sequence is one of the most common features listed as an integral feature of narrative.[8] As a quintessential example, we expect fairy tales to start with some variation of "once upon a time," contain rising action, climax, and falling action, and then conclude with "happily ever after," all taking place one right after the other. The presence of sequential events constitutes a story's flow through time, whether those events are ordered or jumbled.

In the previously discussed example of *mother!*, linear temporality is complicated because the story starts at the end of the previous mother's story and ends with the beginning of the next mother's story, creating a cyclical narrative. In the film *Pulp Fiction*, many simultaneous threads appear disconnected and out of order until the narrative's end. The film *Memento* and the play *Betrayal* are stories told in reverse, so they start with the conclusion of the plot and work backward scene by scene. In these stories, sequence is purposefully adjusted "to maintain a sense of mystery, to keep tension high, and to keep the audience interested."[9] For entertainment narratives, the disjointed sequential narrative can be particularly appealing for introducing a sense of surprise and novelty. Although the narratives are out of sequence, there are clear connections and interrelationships between events, so the sequence wedge of *mother!*, *Pulp Fiction*, *Memento*, and *Betrayal* would be mapped on the meso-ring.

A story that completely lacks causal, ordered events, may be, for some, like philosopher Galen Strawson, not much of a story at all.[10] On a narrative web, we would map seemingly unconnected events on the macro-ring. For example, narratives that are "picaresque or postmodern" may not contain a causally linked sequence of events but are still narratives nonetheless.[11] Narratives that jump in time or construct unclear relationships between the past, present, and future would also be considered meso-ring or macro-ring placements, depending on the relative abstractness of the ordering. The connection between events is mediated by time. While unable to draw a direct causal connection between a particular moment in the past and now, a series of events over time could reflect an overarching story of history.

The micro-ring of sequence would indicate narratives that have closure, which is a story's ability to wrap up the tensions it set in the beginning with a satisfying ending. Kenneth Burke argued that parts of a story should naturally flow from one another, where earlier parts set up audience expectations that are eventually fulfilled by the subsequent parts.[12] A story's order may also be called its "arrangement," which is one of the canons of rhetoric.[13] The arrangement of a story is its logical unfolding to build to an ending based on its beginning and subsequent actions. As religious studies scholar Wilhelm Wuellner defined it, arrangement "is the ordering of substance," or the content of the discourse, "for the purpose of serving . . . the discourse's aim."[14] Sequence is thus linked to the content and action wedges in terms of detailing the order that topics and events/actions unfold within the narrative.

Burke argued that the order of discourse resonates within audiences insofar as it is a recognizable form that conforms to audience's expectations. In other words, the "internal arrangements" of a discourse may follow typical patterns and thus be fulfilling or satisfying to audiences.[15] We expect stories to have beginnings, middles, and endings and certain genres of films to have particular components, key plot points, and tones. Horror films rarely have happy endings, whereas romantic comedies typically do. A cliffhanger may be good for building tension, but audiences typically want that tension resolved and the story eventually finished.[16] Locating different forms, for Burke, details "the varied ways in which [people] seek by symbolic means to make themselves at home in social tensions."[17] With a complete form, we have audience gratification and comfort with expectations met. Consequently, we would map a story that does not reach closure (e.g., is ongoing or unclear) on the meso-ring or macro-ring of sequence as it interrupts the fulfilment of audience expectations based on the form created by the narrative's previous parts.

While not always the case, I found science's stories were more likely to fall on the micro-ring of the sequence wedge than on other wedges. Recounting scientific experiments or discoveries tends to work from start to finish, such as detailing the chronology of an experiment, discovery, or Earth's history. Wuellner noted that "arrangement" was "the earliest framework for technical rhetoric" due to its educational function of building knowledge from previous knowledge.[18] Traditional approaches to science communication rarely involve flashbacks, disjointed sequences, or abstract appeals to time that might warrant a meso-ring or macro-ring mapping. That is not to say, however, that all scientific stories have micro-ring placements. As I will discuss, science's stories can have meso-ring and macro-ring sequences when uncertainty is introduced into the narrative and when there is misalignment between past, present, and future actions, resulting in temporal jumping.

Similar to the character and action wedges, micro-ring sequence placements are not always desirable. For example, environmental rhetorician Phaedra Pezzullo questioned the validity of simple, closed narratives to capture the story of ongoing waste pollution in Warren County, North Carolina.[19] Warren County was seen by many as the birthplace of the environmental justice movement due to its protests of the toxic waste site. However, narratives about the protests labeled them as "successful," thereby "defer[ing] any steps towards change" regarding the landfill's continued toxification.[20] Prematurely declaring endings may limit the ability to act, which connects sequence to action, and may call into question the credibility of a story with changing conclusions, which connects sequence to storyteller.

Sequence, like the other wedges, is subject to audience, situational, and cultural adaptation. While Western stories may be more linear in their arrangements, Indigenous storytelling may incorporate repetition and view time as cyclical.[21] In other words, what may be considered nonsequential and less compelling for Western audiences may resonate and be engaging for Indigenous ones.[22] Therefore, a story's mapping on the narrative web does not, in and of itself, tell us whether the story will be accepted or have high narrative rationality for audiences. Instead, the mapping of sequence provides information about the story's features and level of specificity. From there, we can evaluate a web's constellation and how each feature relates to the others as a coherent whole.

THE SCOPE WEDGE

The scope wedge reflects the breadth of a story, what it covers, and what it does not cover in reference to time and space. Increasing the scope of the story means that it spans more time, distance, events, and circumstances. Science's stories can push at the limits of our scalar understanding. In narrative web terms, science's stories may operate at generalizable levels and therefore be mapped on the meso-ring or macro-ring of the scope wedge, thereby containing considerable amounts of content, time, and space in a single narrative.

Burke called the way one draws the bounds of a story its "circumference," where changing the circumference changes how we interpret the action of a story.[23] Circumferences change the reference point for a story and may trivialize or magnify particular actions or events. For example, if we tell a story with a small circumference where someone killed someone else at a specific time and location, we may conclude that they are guilty of murder. But, if we expand the circumference and view the same action of killing as part of a war between nation-states, the character who at first was labeled a murderer could be relabeled as a war hero.

A story that is so expansive, broad, and abstract to include temporal and geographic frames beyond humanity's direct experiences would be mapped on the macro-ring, while a specific, timely, and narrowed scope that attends to "the particularity of time, place, person, and event" would be mapped at the micro-ring.[24] Much research has been done on the potential benefits of locally situated and place-based narratives to increase audience engagement with scientific topics. Communication scholar José

Castro-Sotomayor proposed the concept of "emplaced climate change" to draw attention to how climate disruption is felt by "people's places and bodies" at a local level, requiring attention to specific cultural and language practices that are hard to translate to broader audiences.[25]

In an experiment, Rainer Romero-Canyas and colleagues, from environmental psychology and public policy, found evidence that exposure to a campaign that emphasized the temporal, geographic, and social proximal effects of climate change led to increased belief in global warming, belief in the scientific consensus, and concern for climate impacts.[26] Instead of being particular or locally meaningful, a story with a broad scope may be hard to follow and find resonance with because of its breadth of information and implications. As Shawn Wilson succinctly argued, "By reducing the space between things, we are strengthening the relationship that they share."[27] Working from smaller scopes, therefore, can emphasize interconnections and proximity to foster an ethics of care and responsibility.

Such conclusions may not always hold true for every circumstance and every story. Broader scopes can be engaging as well as accurate to scientific information. For example, a broad scope can capture the long-term effects of climate change or the billions of years of Earth's history more completely. Anthropologist Misia Landau argued that the study of human evolution works from "descending order of taxonomic scope and of time scale," which "fit together in a coherent epic account of our world."[28] To narrow the scope of a story, such as evolution, to a particular family or community would ignore part of what makes scientific topics so important—their far-reaching and long-lasting implications.

Essentializing stories only to specific, local scenes may also unwittingly cut out historical and genealogical frames that are essential to Indigenous story work and influence those specific, local, and present understandings. While we may think that there is a trade-off between small and large scopes, some stories effectively incorporate both. Indigenous Hawaiian stories, for example, can span large stretches of time and multiple generations as a way of grounding traditional knowledge and honoring ancestors in the present.[29] It is thus possible to have a story mapped across the rings that has a broad scope in terms of its influences and implications while still being grounded in a specific context, as we saw in Brandenburg's narrative web mapped in the character and action wedges chapter.

The following sections outline three themes in institutional scientific stories that are potential barriers to the sequence and scope wedges being mapped on the micro-ring. The three themes are (1) comprehensibility, (2) perceptibility, and (3) uncertainty. These themes will vary in their relevance to each controversy because they span different temporal and spatial scales and relationships between events. Collectively, these themes serve as starting points for considering the role that sequence and scope play in scientific stories and how we may adapt them to public audiences. Although the sequence and scope wedges do not have to be hubs for any given story, outlining the ways scientific stories tend to use these features illuminates potential opportunities for experimenting with new story forms, especially in the face of rival stories that tend to use micro-ring sequences and scopes.

COMPREHENSIBILITY

If the primary purpose of science communication is to instruct, science communicators must attend to how easily audiences can comprehend the story and the information therein. Stories that are incomprehensible, stretch the limits of our imaginations, or are outside the scope of our understanding may fail to resonate with audiences. This section focuses on comprehensibility in terms of the grand scales of time and space within science's stories. I have split the topic of comprehensibility into time and space for convenience, but the topics often intersect and compound one another in science communication. By attending to them separately, we can consider how scientific controversies have different dynamics within temporal and spatial comprehensibility. As we will see in the following analysis, climate change and evolution's stories feature most prominently in temporal comprehensibility, while vaccination and COVID-19's stories are more prominent in spatial comprehensibility.

Time

In 2019, the Anthropocene Working Group, made up of thirty-four interdisciplinary scientists, voted to propose that the Anthropocene, the era of humans' most profound effect on the Earth, was a new geological epoch.[30] If the Anthropocene is approved as a new time period, the prior Holocene, which lasted for approximately 11,700 years, will be over. The Holocene's nearly 12,000 years may seem like a long epoch, but it is relatively short. The epoch before the Holocene, the Pleistocene, lasted for more than 2.5 million years. In telling the stories of climate change, discourses of the Anthropocene and geological time are becoming more common. As EarthHow noted, "The geological time scale of Earth is *almost unimaginable* to us," because our conceptualizations of time are miniscule compared to the vast time frames of the universe.[31] The National Academy of Science's report on climate change contextualizes a graph of Earth's annual average temperatures by noting that it describes "land and ocean measurements from 1850 to 2012."[32] The American Meteorological Society described the time frame of climate change as "decadal to multi-decadal timescales."[33] Our perceptions of time in the stories we tell about climate change challenge us to comprehend timescales both backward and forward.

Multiple studies have found evidence that people have a limited ability to conceptualize more than ten to twenty years into the future, especially relating to climate change impacts.[34] While a span of a few decades to a few hundred years may not seem very long, especially given the scale of most scientific disciplines, it is still typically outside of people's direct experience, affecting how much the stories of climate change resonates with them. Anthropologist Peter Rudiak-Gould argued that being able to sense, feel, and understand climate change is made difficult due to its "arcane mathematics, high-tech measuring devices, and inhumanly large temporal and spatial scopes."[35] Concerned primarily with our own immediate well-being, people may find themselves unmotivated to prioritize environmental policies over issues of safety, economics, and education that may seem more pressing and consequential.[36] Sabine Pahl and colleagues, a team of psychologists and social scientists, argue that the "standard timelines (e.g., 2050, 2100) used by climate scientists are therefore *not meaningful* to the general public."[37] I have italicized "not meaningful" in the previous quotation to emphasize the importance of meaning within storytelling. To return briefly to Walter Fisher, he argued that stories are "words and/or deeds" that have "meaning for those who live, create, and interpret them."[38] Without meaning, climate predictions, by Fisher's definition, might fail to constitute a reasonable narrative that audiences may adopt. In narrative web terminology, such a story would have an abstract, macro-ring scope that stretches into the future over many years, potentially emphasizing temporal distance between audiences and climate change's impacts.

Climate change's narratives necessarily engage large temporal scales, but they pale in comparison to the time frame of evolutionary science. Descriptions of evolution often start with the origin of life or even the Earth itself. The National Academy of Sciences noted "there is evidence that bacteria-like organisms lived on Earth 3.5 billion years ago, and they may have existed even earlier, when the first solid crust formed, almost 4 billion years ago."[39] Karl Giberson, a physicist and writer on the creation-evolution controversy, characterized evolution's timescale thusly:

> Most of us have trouble conceptualizing 500 years, what about 5,000, 5 million, 5 billion years? We can't imagine what that is at all. We can't imagine the size of the universe, the number of stars in our galaxy, and the number of galaxies. All of these things push our imaginations.[40]

If decades and hundreds of years are hard to understand, then millions and billions operate well outside general human comprehension. The National Research Council argued that "evolution is the most difficult set of concepts to teach" because "it involves complex biological mechanisms and time periods far beyond human experience."[41] Communicating science to public audiences, including students, is difficult in the face of these large timescales.

Even a time span of thousands of years may be out of reach for many people to comprehend. For example, the National Research Council argued that people often have a "hard time understanding the difference between a thousand and a million, much less between a million and a billion."[42] I located an example of this difficulty in an erratum to an article published by *National Geographic* that discussed the likelihood of various human origin stories. The erratum notes,

> Two corrections have been made to this article. In the first sentence "million" has been changed to "billion." In the seventh paragraph, "10 followed by 2,680 zeros" has been changed to "1 followed by 2,680 zeros." Many thanks to readers for pointing out these typos.[43]

The use of the term *typo* implies that the original article had made minor errors in recounting the results of the scientific study. On first glance, the error of replacing the accurate 3.5 billion with 3.5 million may seem like a simple mistake of a single letter. However, this change indicates a thousand-fold difference in the age of the single-celled organism, which is an error equivalent to reporting that the average woman's height is fifty-four hundred feet, or more than a mile, tall, instead of five feet four inches. This example points to the large scopes within evolution's stories and difficulties in conceptualizing them in meaningful ways.

To capture evolutionary timescales, visuals can be helpful. Visual representations of evolutionary time tend to modify or distort a normal linear time line because of the immense amount of time being represented. For example, some representations use a spiral instead of a straight line, pushing the past deep into the distance to make room for a sliver of human presence at the front. Visual choices must be made to keep the full scope of evolution in the same picture while providing contemporary details that are relatively trivial given the full image. In the spiral map image (see figure 5), the time line goes back to the Earth forming approximately 4.5 billion years ago and is detailed enough to just show the origins of humanity and the start of industrialization in the Quaternary Period. This visual representation highlights the difficulties of representing relevant and contemporary information on accurate timescales.

The long history of evolution is a testament to science's ability to study the past. However, with a lack of understanding or at least a way to make sense of millions and billions of years in the context of our lives, evolution has low narrative fidelity, meaning that it does not resonate with the timeliness and immediacy of our existence. Alternatively, creationism's chronology, as told by the Bible, has a more accessible time span; creationists place the age of the Earth around six thousand years, based on the genealogical information present in the Bible.[44] This scope puts less time between contemporary life and its divine origins, making creationism's stories easier to comprehend and more narratively probable.

Shrinking temporal scopes within science's stories can happen in two ways. First, science communicators may simply truncate the scope of time being discussed to make climate change more temporally present and proximate. Alternatively, temporal scopes can be symbolically adjusted to

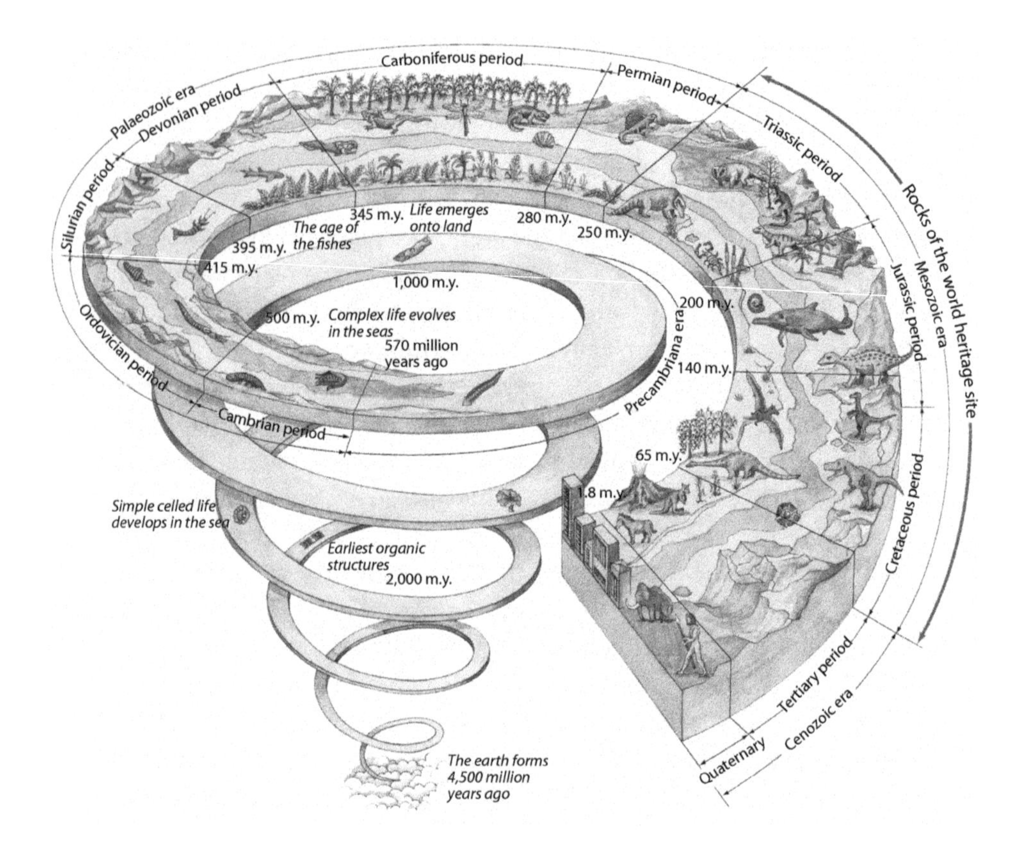

Figure 5. An image of evolutionary timescales using a spiral instead of linear shape to capture the vast lengths of time in Earth's history. Note that the spiral starts well into the distance and human figures do not appear until the very last section of the spiral. Reprinted with permission from the Jurassic Coast Trust.

present large time frames in more comprehensible terms. A science communicator, for example, could contextualize multidecadal timescales through discussing lifetimes. Noting that climate change's effects will happen "in our children's lifetimes" or "by the time someone born today graduates high school" can put temporal frames into more manageable time frames that follow people's direct experiences. Similarly, past data could be connected to historical points in time. For example, the National Academy of Science's timescale of 1850 to 2012 could be reframed as between the thirteenth and forty-fourth presidents (Millard Fillmore to

Barack Obama). These changes can represent scientific information in more concrete terms despite their large temporal scopes.

Space

Large scopes occur not only temporally but also spatially in science's stories. Rudiak-Gould characterized the physical sciences writ large as working from aggregations that necessarily operate at large scales. One implication of aggregate work is privileging the global and universal as "objective fact" and relegating the local to "subjective experience."[45] In searching for objectivity, institutional scientists tend to abstract upward, encompassing large enough data points to confidently report aggregate changes. Rudiak-Gould rightfully points out that this orientation to knowledge is not natural but is a construction of a particular "political outlook,"[46] such as a norm within scientific institutions guided by a Western approach to knowledge. There is, of course, great power in the local, a particular type of experiential authority that can be found in the wisdom of ancestral knowledge, community-based practices, and regional observation. Adjusting the scope of science's stories to recognize local knowledges and emphasize community scales, which are often most important to audiences, can be a way to counter the typical immensity of scientific narratives' spatial scopes.

Climate change's stories, understandably, have a global scope that calls for "substantial and sustained global mitigation" and "international cooperation."[47] The fifth report from the Intergovernmental Panel on Climate Change acknowledged that global cooperation is difficult and varied, noting, "Existing and proposed international climate change cooperation arrangements vary in their focus and degree of centralization and coordination."[48] While difficult, the Intergovernmental Panel on Climate Change argues that implementation "across geographical regions" in a concerted, orchestrated effort will make climate policies "more effective."[49] Focusing on global causes and effects means an expanded scope that may put literal and figurative distance between a person and their interest in climate change.[50] Geographer Dylan Harris argued that traditional stories tend to emphasize the human element, but "the climate story precedes and exceeds us."[51] Indeed, the term *global warming* only makes sense on global scales where average temperatures show increases over time amid cycles of warming and cooling. However, such a large scope may overshadow regional differences and how the burdens of climate change are not evenly distributed. A global scope, in other words, may unfairly universalize climate change as manifesting in the same ways everywhere. However, climate change amplifies weather patterns and makes extreme weather more common, so it is not unfeasible that our future may look like both *Water World* and *Mad Max: Fury Road* in different regions.

Stories about vaccination tend to focus on issues of herd immunity and community protection as opposed to individual children, thereby widening the circumference of vaccination's stories from families to nations and even the world. This expanded scope justifies a collective approach to health where individual needs are seen as part of the public good. While parents are primarily concerned about their individual child's well-being, science's stories of vaccination sometimes span large groups of people. The American Academy of Pediatrics argued that "immunization coverage, access, and affordability are issues that go beyond US borders" and quantifies the number of underimmunized as "19.4 million infants worldwide."[52] This website notes that "more efforts are needed to increase immunization coverage, access, and affordability, and to fight against vaccine distrust globally, especially as infectious diseases know no boundaries."[53] While people are often primarily concerned about their personal and local conditions, these statements frame vaccination as a global endeavor, potentially expanding the scope beyond people's immediate frame of reference.

On its website listing common ingredients in vaccines, the Centers for Disease Control reports that there are "one billion doses of vaccines manufactured worldwide each year," emphasizing a large and perhaps incomprehensible amount of vaccine dosages needed to keep the population safe.[54] The World Health Organization also, understandably, emphasizes global vaccination rates and preventable deaths. The World Health Organization noted, "Each year, vaccines prevent more than 2.5 million child deaths globally. An additional 2 million child deaths could be prevented each year through immunization with currently available vaccines."[55] On the one hand, these numbers celebrate the great medical advancements of vaccination. On the other hand, they engage a large, global scope instead of focusing on individuals' or families' decision-making about vaccination for themselves. One contributor to VaxTruth, an anti-vaccination website, wrote, "The death or lifelong disability of your child is deemed by [medical experts] as acceptable collateral damage" in support of the "greater good."[56] Although not intended by medical organizations, vaccine skeptics may interpret a community- or global-level focus as superseding considerations of individual health decisions.

Vaccination discourse also emphasizes herd immunity, or vaccinating people to reach a threshold where those who could not otherwise be vaccinated are protected. The Centers for Disease Control argues, "Vaccination is important because it not only protects the person who gets the vaccine, but also helps to keep diseases from spreading to others, like family members, neighbors, classmates, and other members of your communities."[57] Although this statement includes protection for the child or person vaccinated, the scope frames others in an increasingly broad circumference as also impacted by people's vaccine decisions. Family Doctor argued, "Because these people [very young, immunocompromised, the elderly] can't be vaccinated, it's very important everyone else gets vaccinated. This helps preserve 'herd immunity' for the vast majority of people."[58] *Herd immunity*, as a term, emphasizes the herd or group as opposed to an individual.

There are additional complications with the term *herd immunity*, which connotes animals and livestock being vaccinated. Scientists developed the term to refer to the "herd" of mice being tested for vaccine efficacy.[59] Akin to anti-vaccination arguments that classify vaccine advocates as "sheep" who uncritically follow government recommendations, messages about herds may undermine people's feelings of autonomy and agency. Anti-vaccine advocates frequently denounce "herd immunity" as compromising personal safety and decision-making.

A VaxTruth web page argued that the government is focused on the national level, so is not concerned "about whether or not your child might be more vulnerable to vaccine-injury." The web page continued, "The government has made the decision that [it] is most important to protect the general public from infectious diseases," so it is up to parents to decide "whether or not vaccines are advisable or safe for your particular child."[60] Vaccine hesitant and skeptical groups redraw the circumference to focus on individual children and parental decision-making, against the scientific stories' expansive scopes. One parent, recounting her son's story on VaxTruth, noted, "He's not a statistic. He's a person," highlighting the dichotomy she feels between the aggregate lens through which scientists and doctors look at unvaccinated children and the individual lens through which she views her child.[61] Whereas science's stories may emphasize the community-level impact of widespread vaccination, parents may resonate more with a story with a narrowed circumference that justifies a focus on their own household.

As with all the wedges, we must not privilege only the narrow and local. Appealing to global impacts can be an effective way to motivate action on scientific topics by elevating a problem's importance and impact.[62] Large scales can invoke a sense of "wonder," which can motivate a public's attention by ascribing praise and value to a scientific topic.[63] Rudiak-Gould suggests that large scales could be communicated in plainer language, such as by substituting "global" with "pervasive" or "multisited" to emphasize specific, prolific instances of change and using "long-term weather change" instead of "climate change" to relate one's personal experiences of weather to the abstract phenomenon of changing climates.[64] Additionally, large and small scales can be put together to emphasize individuals as part of communities. For example, the World Health Organization argued that "vaccines have an expansive reach: they protect individuals, communities, and entire populations."[65] Through attending to narrative features, we can make strategic choices to make spatial scales more comprehensible and relatable to audiences.

PERCEPTIBILITY

Even if audiences are able to comprehend science's scales, they may only be able to imagine them as opposed to being able to perceive them directly. By perceptibility, I refer to the ability of science's stories to be sensed and felt by audiences. Perception of scientific controversies may come in physical forms, such as breathing in smoke from wildfires or firsthand experience of a communicable disease. Additionally, we can perceive scientific controversies through the symbolic representation of their relative temporal and spatial proximity. Considering that people are most often concerned with what they can perceive in their "direct environment," we are "geared to prioritize short-term consequences of behavior and immediate futures" and "close cause-effect relationships" rather than long-term or delayed ones.[66]

Similar to how it can be hard to understand vast quantities of time and space, incredibly small scopes may also defy understanding in being mostly invisible to our natural senses. For example, our bodies are crawling with microscopic organisms that live on our dead skin cells, burrow into our eyelashes, and crawl through our scalp, but we are often unaware of them unless they cause a larger, more noticeable issue such as an infection. We characterize an eyelash or a fly or a microscopic bacterium as "small" in relation to ourselves; scale is a measurement of discernable difference between objects.[67] To us, a fly may be small, but to a proteobacterium, a microorganism that colonizes flies' guts, a fly is its entire world.

Although they technically span narrower scopes, stories with microscopic and invisible scales are mapped on the meso-ring of the narrative web. This placement indicates the lack of perceptibility of small scales, which are abstract and complex. In other words, small scales do not necessarily mean that something specific is being talked about; many small-scale narratives are implicitly abstract. Genes, carbon dioxide, and viruses are both "global and microscopic" in that they are "simultaneously too small and too large to be directly" experienced.[68] Small scales must also be put into context and made relevant to audiences in a frame that they can understand and sense as meaningful.

As previously mentioned, climate change's scope involves large periods of time. While current and predicted impacts are large, how we communicate these changes quantitatively is not, on first look, impressive. Using a rhetorical question, NASA wrote, "So, the Earth's average temperature has increased about 2 degrees Fahrenheit during the 20th century. What's the big deal? Two degrees may sound like a small amount, but it's an unusual event in our planet's recent history."[69] Referencing a historical time scope, NASA argues that what seems like a relatively minor change of a few degrees becomes noteworthy when compared to our past. In doing so, they acknowledge that two degrees, on its own, seems like an insignificant change. The National Academy of Sciences notes that "Earth's average temperature has increased by about 0.8°C (1.4°F) since 1900."[70] The difference between one degree is hard to perceive, so asking audiences to make sense of a yearly change of 0.007°C (0.01°F) is far-fetched indeed. Depending on where they live, people may be personally familiar with temperature swings from 0°C to 42°C (32°F to 107°F) in a calendar year, so statements of a few degrees globally may seem insignificant.[71]

One way that people make sense of the climate is through the palpable feelings of weather such as changing temperatures, even though much scientific discourse cautions against conflating the two. A report by the American Meteorological Society argued that people should not rely on yearly changes that they experience as a reliable marker of climate change:

> Global surface temperature fluctuates from year to year and decade to decade from interactions of the atmosphere, ocean, land surface, and cryosphere (the snow and ice-covered portions of Earth); these processes represent natural climate variability. Consequently, not every year is warmer than the preceding one, globally.[72]

The American Meteorological Society asks people not to trust their personal experiences but instead to think about multiple decades at a time. A similar argument was made by the Environmental Protection Agency, which noted that people should not expect "a smooth pattern of steady change" because "the Earth is a complex system, and there will always be natural variations from one year to the next—for example, a very warm year followed by a colder year."[73] Our senses and direct perceptions of climate change via weather are thus discarded in favor of longer term thinking where changes are only meaningful on less tangible aggregate levels.

Rival stories of climate skepticism, however, emphasize personal experiences of weather changes. Take, for example, a tweet from former US president Donald Trump after a cold front:

> In the East, it could be the COLDEST New Year's Eve on record. Perhaps we could use a little bit of that good old Global Warming that our Country, but not other countries, was going to pay TRILLIONS OF DOLLARS to protect against. Bundle up![74]

In this tweet, one's physical perception of temperature was used as evidence that there is no global warming. Making a similar argument, Senator James Inhofe (R-OK) brought a snowball to the floor of Congress in 2015 to imply that the presence of snow contradicts evidence of warming. While farcical on its face, Inhofe's stunt appeals to our direct experience of the weather and our environment, which are more present and pressing than long-term, large-scale changes. Many aspects of climate change are so small as to be imperceptible. For example, carbon dioxide and "greenhouse gases" operate on molecular, "microscopic" scales that are too small "to be directly seen" and yet have global consequences.[75]

The inability to "perceive" climate change is, of course, a privileged stance. Countries and communities around the world are already feeling the impacts of climate change in the forms of food insecurity, changing weather patterns, and climate-induced or climate-worsened disasters. Arguing that climate change is "visible" based on one's "spatial or temporal referent," Rudiak-Gould noted, "Climate change is invisible to sheltered, ecologically naïve urban Westerners but visible to ecologically savvy indigenous people, nature workers, and frontline communities."[76] The groups most affected by climate change are least responsible for rising emissions and industrialization. Climate stories can help emphasize climate injustices by using scopes both to illustrate the importance of present-day action to influence the future and to locate consequences already happening.

Providing examples is a potentially useful strategy to pair with scientifically accurate understandings of our climate systems and predictions. In *An Inconvenient Truth* and subsequent public talks, former US vice president Al Gore uses a series of examples to emphasize the death toll and economic impacts of global warming. In quick succession, Gore uses the rhetorical strategy of *copia,* defined as the quantity of information, as evidence itself of climate change's impact. Providing environmental examples can also help break down audiences' misconceptions about climate change and help them make further connections between environmental changes and their daily lives.[77]

Evolution's stories involve small scales in terms of gene variations and mutations that add up to larger changes over time. The National Research Council described how evolution's processes happen "at the microscopic scale" where life is made up of "invisible but powerful chemical force[s]."[78] The Understanding Evolution website notes that the definition of evolution "encompasses small-scale evolution" on the level of DNA, genes, and alleles. These components of life are infinitesimally small, so are made meaningful by describing their bountiful presence. In *One Plus One Equals*

One, biochemist John Archibald noted that "ten trillion [is] the minimum number of cells comprising a human being."[79] Microscopic interactions are not readily observable, especially ones happening inside of bodies. Rival storytellers to evolution characterize the diversity of life as "mind-boggling" and "mind-staggering" to indicate how much variation evolution's stories say has occurred due to these minute genetic-level changes.[80]

Although these changes are apparent with the passage of time, some changes at the genetic level that constitute an evolution of species may not manifest in any visual or physical way. In other words, changes in life-forms happen over hundreds of years of which any example is a snapshot in time. We cannot see evolution in action because it is only identifiable through the aggregation of small changes over long periods. For example, while biologists measure alleles, genetic frequencies, and interpopulation breeding to denote evolutionary change, paleontologists measure such changes through fossil records.[81] None of these scientific measurements are readily accessible or perceivable to audiences without specialized training.

Charles Darwin even acknowledged that the changes of evolution are so small that humans are not well-equipped to perceive them. He noted,

> We see nothing of these slow changes in progress, until the hand of time has marked the long lapse of ages, and then so imperfect is our view into long past geological ages, that we only see that the forms of life are now different from what they formerly were.[82]

The Smithsonian Institution's Human Origins Program echoed this difficulty to experience evolution by noting "evolution doesn't happen all at once,"[83] and, therefore, it is difficult to perceive through direct experience.

Similar to climate change, the language used to discuss evolutionary timescales can complicate our ability to perceive evolution as temporally relevant. For example, many scientific texts use billions of years, but the spiral map previously discussed in the "Comprehensibility" section lists 4.5 billion years as 4,500 million years. The use of thousands of millions instead of billions appeared frequently in my research and is likely a holdover from different numbering conventions between countries.[84]

To give an impression of the various numbers being discussed in evolution's stories, and thus the numbers that the public must sift through when reading these stories, I have assembled the following list of dates and descriptions: "rocks older than 3.5 billion years have been found on all the Earth's continents"; "layered rocks known as stromatolites that appear to have resulted from the actions of bacteria at least 3.4 billion years ago"; "sedimentary rocks that are between 540 million and 635 million years old"; "the existence of wormlike creatures as long as 1 billion years ago"; "they are 505 million, 385 million, and 340 million years old, respectively"; "simple microbes fused together more than 2.5 billion years old"; "we do share a common ape ancestor with chimpanzees [that] lived between 8 and 6 million years ago"; and "these estimates yield 4.54 billion years as the age of Earth and of meteorites."[85]

Without context or a point of comparison for these numbers, it is difficult to perceive them as markedly different beyond that they are all describing long lengths of time. Understanding the relative distance between these stages of time is dependent on one's ability to discern meaningful differences between hundreds of thousands and millions of years. For people who live and experience a time frame of one hundred years, perceiving longer timescales is a difficult shift.[86]

Alternatively, creationism's stories emphasize the importance of the senses and direct perception as ways of knowing.[87] Answers in Genesis CEO Ken Ham argued that regarding science's dating measurements of the Earth, "You weren't there, you can't observe that."[88] In arguing that observation of the past can only come from a primary source, Ham notes, "There is only one infallible dating method. It's the witness who was there."[89] Doing science is framed as simply trusting your senses and observing the world around you. In this way, creationists make a direct appeal to people's perceptions in lieu of providing traditional scientific evidence.[90]

Science's COVID-19 narrative, especially in its early stages, also appealed to the senses, or the lack thereof. Organizations like the Centers for Disease Control emphasized personal symptoms such as a loss of taste and smell, a heavy fever, fatigue, and other COVID-19 side effects to encourage people to wear masks and get vaccinated. An emphasis on "sensing" COVID-19 gave some people the impression that they were not at risk; not knowing anyone who had gotten sick or had only mild symptoms fortified vaccine hesitation and considerations of the vaccine as unnecessary. Attempts to contain the virus work counterintuitively by providing less evidence that the virus is something dangerous and worthy of preventative action to contain. Similar arguments are made against other vaccines, where temporal distance from the effects of such diseases as measles and polio contribute to vaccine hesitancy. Medical and health researchers Daniel Salmon and colleagues called this tension the "familiar versus exotic" and argued that vaccine hesitancy is in part caused by people being "more familiar with common health problems that are alleged (without scientific support) to be caused by vaccines, like autism, than diseases they are not familiar with, such as polio, measles, and diphtheria."[91] Similarly, asking people to mask up during COVID-19 offers people the potentially uncomfortable experience of wearing a mask against the unknown, potential symptoms of COVID-19.

Technical experts can often understand information that most members of the public cannot because the latter have not been trained to understand such large and small scales. This is not to say that it is impossible to develop an understanding of incredibly large or incredibly small scales, but it often comes with scientific training, personal interest, or occupational necessity. Science communicators can play an important role in helping public audiences develop this appreciation and understanding of scale.

UNCERTAINTY

With more tools and methods at scientists' disposal, we might assume that there would be more knowledge and less uncertainty. However, additional information creates more variables and more questions, which has been described as the "multiplication of uncertainties" for the modern scientist.[92] Uncertainty is a natural component of the scientific process; few facts are proven to the level of 100 percent certainty, such as gravity, and are expressed in terms of probabilities (i.e., *p*-values).[93] Scientific methods attempt to reduce uncertainty as much as possible, but there is still always a measure of uncertainty in the processes of measuring, interpreting, generalizing, and predicting. Uncertainty emerges in science's stories through predictions, misalignment with expectations, and unclear endings.

One of the strengths of scientific inquiry is its predictive power. We run experiments and collect data to know more about the world and its functioning, but also to prepare for and understand future events. Even though circumstances are always changing, scientists can try to predict future effects and possible outcomes. Rhetorician of science Lynda Walsh argued that predictions are contested arenas for science communication because they denote things not-yet-happened by a discipline that understands and measures empirical reality.[94] Despite these differences between current measurement and future occurrences, scientists have long evoked what Walsh calls the "prophetic ethos," whereby they warn and advise publics on prudent policy decisions based on privileged knowledge from their scientific training and experimentation.

Advising publics is a natural by-product of the scientific process because information often prompts the need for action in response.[95] If scientists measure the likelihood of an earthquake and report that it is incredibly likely within a certain amount of time, these statements of fact signal a necessary and urgent course of action. Many public audiences, however, do not view science's proper realm to be advising, and scholars have been wary of a ruling technocratic elite that could overshadow public decision-making.[96] Walsh referred to this constraint on science communicators' ability to be prophets or advise on policy as the is/ought boundary.[97] Advising what we "ought" to do based on what we model the "is" to be in the future introduces multiple levels of uncertainty into scientific predictions. When uncertainty is part of the narrative, sequences of cause and effect between interrelated events can be disturbed, thereby creating room for skepticism and potentially decreasing audience adoption of the narrative.

While previous sections have discussed climate change's timescales, this section attends to statements that introduce uncertainty between the present and the future. The US National Research Council wrote about the impacts of our actions by noting "emissions reductions choices made today matter in determining impacts experienced not just over the next few decades, but in the coming centuries and millennia."[98] For some, this statement might elevate the importance and urgency of present-day action. For others, this statement could be interpreted as overstretching science's predictive power, for how can we know what will happen thousands of years from now?

Part of scientific research is the acknowledgment of its inherent uncertainties. It is scientifically accurate to hedge or qualify predictions, but the inclusion of uncertainty in scientific texts may also provide ammunition for climate skeptics. The National Academy of Sciences noted that "it remains difficult to predict the details" of warming in various regions using current climate models.[99] The American Physical Society noted that "the magnitudes of future effects are uncertain" and that "scientific challenges remain in our abilities to observe, interpret, and project climate changes."[100] NASA noted that climate change consequences "are difficult to predict," but notes that some effects are "likely."[101] The American Meteorological Society acknowledged these challenges as well, noting that while models are "based largely on fundamental physical laws and well-understood physical principles," climate models are prone to "model errors" due to how "crucial processes like clouds and convection, ocean eddies, deep water formation, and carbon cycle remain crudely represented" in current models.[102]

Warming predictions are based on tested and reliable scientific methods but are often characterized by climate skeptics as hysterical, alarmist, and apocalyptic. For example, interdisciplinary researchers M. Jimmie Killingsworth and Jacqueline Palmer analyzed how "predictions of environmental disaster" were characterized "as the empty fantasies of confused leftists," and linguist Brigitte Nerlich critiqued the association between climate predictions and "doomsday prophecies."[103] There is perhaps no more iconic example than the cover of Senator Inhofe's book *The Greatest Hoax: How the Global Warming Conspiracy Threatens Your Future,* which shows a man in a suit consulting a crystal ball filled with the Earth. DiCaglio notes that scientific practices have long resisted "the speculative" through their "rigor, skepticism, and discipline,"[104] but this does not stop accusations of science's models as being akin to a fortune teller's predictions as a way to undermine the relationship between current actions, or lack thereof, and projected consequences.[105]

Climate skeptics misconstrue scientific methods in order to cast doubt on climate science's predictive abilities. For example, the Heartland Institute, a climate skeptical think tank, described climate scientist Michael Mann's hockey stick graph as "coddled [cobbled] together from bad data [and] bad computer code." Heartland Institute contributor

H. Sterling Burnett notes that "climate models have inherent, systematic errors which undermine their temperature and other climate projections," concluding that the uncertainty of climate measurements means that "no climate crisis exist[s]."[106] Another climate skeptical think tank, the Acton Institute, argued that contemporary climate science is based on "highly speculative computer climate models."[107] Even if audiences do not fully believe climate skeptical think tanks, these publications circulate in search engine results and social media feeds and introduce uncertainty into climate science's predictive abilities.

In the sense that climate prophets attempt to warn humanity to change their behaviors, one could interpret failed climate predictions as positive signs of forestalling the worst effects or of more finely tuning climate models. Climate skeptical groups, however, use failed predictions as a reason to distrust future predictions. In an article commemorating the fiftieth anniversary of the first Earth Day, the Heartland Institute published an article filled with quotations from scientists predicting a climate disaster that has not (yet) come true. The article ends with the author calling for a "need to critique the 2020 doomsday predictions in the year 2050 and see if they were any better than those from the first Earth Day 50 years ago," implying that they will not be.[108]

The issue of climate models is a double-edged sword in science communication. Explaining details about how we know future climate information may help alleviate concerns about how scientific predictions are made, but these statements also emphasize the uncertainties and variation that occur in those models. I do not advocate completely removing uncertainty from science communication, which would be misleading and unethical. Instead, scientists should acknowledge uncertainty as an obstacle that must be navigated in discussing climate science. Science communication should ease worries and put relative uncertainties into context in terms of the vast amount of information that is known.[109]

The stories of vaccination can also have an uncertain future in terms of weighing risks between the possibility of contracting a disease in the future against perceived risks in the present. Unlike government guidance regarding car seat requirements, vaccination decisions are much more personal and intimate because they prescribe mandates for what is injected directly into bodies. A strap or a car seat is easily removed or adjusted, but a vaccine is a much more permanent decision. The Centers for Disease Control estimates the risk of adverse reactions to vaccines, such as developing encephalitis (i.e., inflammation of the brain), as exceedingly rare, potentially as low as one in one million,[110] but for some parents, that risk is still too high. As VaxTruth, an anti-vaccination advocacy group, explains, "When it happens to your child, the risks are 100 percent."[111] In other words, vaccine hesitant discourses evoke vaccine injuries as a distinct possibility, thereby encouraging delays, modified schedules, or skipped vaccines.

Once a decision to vaccinate has been made, people cannot turn back the clock or reverse a decision, which is why vaccine hesitant discourse encourages parents to "take your time. You are in no rush. Take all the time necessary to learn everything you can so you can make an informed decision."[112] While measles or polio or even COVID-19 may not seem like a real, present, dangerous risk, vaccine injuries may be portrayed as such. On its website, VaxTruth lists over thirty personal stories of children that have suffered "vaccine injury" or died after being vaccinated.[113] These stories foster uncertainty in deciding to vaccinate by providing perceived evidence of vaccines' harms. To draw upon Burke's language of circumference, vaccine skeptical groups like VaxTruth draw the circumference around individual children harmed by vaccines in the present instead of the wider temporal circumference of a child's future health. In other words, it may be easier for people to weigh the risks of what appears to be a present, felt vaccine injury against the potential for disease later in life.

In the case of COVID-19, there was similar discourse about preventative measures that asked for present behavior change to offset a potential but uncertain future. Wearing masks and staying indoors are behaviors that promise an intangible future benefit (*not* getting sick or a reduced likelihood of getting sick) for a present sacrifice or inconvenience. Given the novelty of the virus, many countries around the world instituted lockdowns to try and contain its spread to prevent future overloads on the hospital system if many people were to contract COVID-19 at once. Due to the misalignment between threat perception and response, some people saw lockdowns as overreactions, which introduced uncertainty into scientific and political decision-making.[114]

Later into the pandemic, statements from the Centers for Disease Control included information about future "variants," which occur due to "genetic changes to the virus [that] happen over time."[115] Getting the vaccine early can help provide protection against future variants and may help prevent mutation caused by the virus moving through infected bodies. Writing for the public-facing outlet *The Conversation*, microbial evolution scientist Ed Feil wrote that "predictions about the evolutionary course of the [corona]virus, and specifically changes in virulence, will always be riddled with uncertainty."[116] As more and more variants emerge, people are experiencing "COVID fatigue." COVID fatigue is not a symptom of contracting COVID-19, but the fatigue of continuously monitoring and acting to prevent contraction and transmission of the virus.[117] The regularity of variants emerging, although predicted as a consequence of not achieving herd immunity, has prompted conspiracy narratives that COVID-19 variants are planned releases. On July 29, 2021, conservative pundit Melissa Tate, who authored a book that challenges critical race theory and the notion of White privilege, tweeted what appeared to be a spreadsheet listing the "launch dates" for COVID-19 variants from June 2021 to February 2023.[118] Despite the image being repeatedly debunked as fake, this has not stopped its circulation across social media platforms such as Twitter and Instagram and the perpetuation of uncertainty about the future development of the virus and its source.[119]

Seeing COVID-19 as planned gives a more complete sense of a beginning, middle, and ending to the narrative, instead of an uncertain future with the possibility of more variants and shifting Centers for Disease Control guidelines. The uncertainty surrounding COVID-19's potential ending was further complicated by President Joe Biden's statement on *60 Minutes* in September 2022 that "the pandemic is over."[120] Although people are still contracting COVID-19 and dying from it, Biden's words attempt to create a clear ending to the COVID-19 story.[121] For those still suffering from COVID-19's long-haul effects such as brain fog and shortness of breath, such a statement is premature if not directly harmful.[122] Furthermore, an "end" to the COVID-19 pandemic interrupted the time line regarding the push of the new bivalent vaccine approved in August 2022 that prevents better against the omicron variant. If the pandemic is over, then why is a new booster needed? For Biden, the ending of

COVID-19 may be a necessary closure for the fidelity of the Democrats' political narrative as the November 2022 midterm elections approached. Premature endings may ring false for audiences and fail to recognize the ongoing struggles in the still-unfolding controversy.[123]

Although uncertainty can introduce problems for science's stories, we cannot and should not completely avoid it. For example, stories about the future can be uplifting and motivating. The future symbolizes hope, opportunities for change, and reimagining our current world. People may think more readily in direct experiences and short-term effects, but we should not be afraid to draw in long-term and future effects to contextualize and amplify the importance of scientific topics.

CONCLUSION

The wedges of sequence and scope work together to build the setting for scientific stories. From great breadths to tiny depths and from the far reaches of the past to the distant future, how we draw the circumference of our narratives influences the other features. Considering sequence and scope in tandem with one another can be a powerful tool for science communicators by prompting reflection on the setting of the story and the information contained therein. Without providing specific prescriptions for scientific stories, which will be situation, audience, and purpose dependent, I offer the following questions to brainstorm the sequence and scope wedges for science communicators and storytellers:

- What are the temporal and spatial scopes of the story I'm telling? Are these scopes within a layperson's likely frame of reference or would they require specific training to understand them?
- Have I contextualized a large- or small-scale event or process? What points of comparison are there for more relatable scales?
- Can I incorporate a variety of scope rings by discussing specific examples, community examples, and global examples that build on one another?
- Does my story follow disciplinary conventions of sequence from the past to the present or from a cause to an effect?
- Can I incorporate a variety of sequence rings by moving away from a linear, cause-and-effect model? Would my story benefit from moving

backward, jumping in time, or using techniques of flashbacks and flashforwards?

- Does my story invite audiences to occupy times or spaces other than their present moment or situation? How can these be expressed in relatable terms or experiences?
- What are audiences' expectations for the "form" or order of my story? Does my story align with or violate those expectations? Why and with what purpose?
- What are the inherent uncertainties in my story? How can I accurately represent these uncertainties while still communicating the status of scientific knowledge on the topic?

These questions are meant to spark reflection and contemplation as opposed to prescribing an ideal ring placement for the sequence and scope wedges. Instead of over-relying on scientific norms or uncritically adopting standard patterns of communication, narrative webs invite us to consider each narrative feature as a rhetorical decision to emphasize, minimize, select, deflect, include, or exclude.

The stories of climate change, evolution, vaccination, and COVID-19 involve different deployments of space and time to capture each topic. From the microscopic functioning of viruses in the body to the billions of years of Earth's history, science's stories incorporate mystifying scales outside of our daily experiences. Issues of sequence and scope are not necessarily unique to scientific topics. Many academic disciplines encompass long temporal or geographic scales or introduce uncertainty in predicting future events, such as economics, demography, history, among others. Other disciplines may also find it useful to pay attention to sequence and scope when speaking to public or other nontechnical audiences.

We do not want to foster additional wariness toward large numbers, time spans, and distances by cutting them completely from science's stories. Much of climate change's stories is based on the dramatic impacts that humans and industrial pollution have caused on the Earth in extraordinarily little time compared to the many years before. The same is true for evolution, where one must have a good sense of the time frame in which evolution has occurred to understand its intergenerational mechanisms. In terms of spatial scopes, climate change is defined by its global aggregate changes, and thinking globally is necessary to celebrate the many lives that vaccinations have saved.

Similar to the character and action wedges, we do not want to privilege the micro-ring of the narrative web at the expense of scientific accuracy. Only looking at the most recent part of history and narrow spatial scopes would prematurely sever scientific stories and thereby inhibit complete or deep understanding. Narrow conceptions of sequence that privilege the purely chronological may deter more creative storytelling techniques like flashbacks, flashforwards, and building suspense. A focus on sequence and scope encourages us to consider ways to foster understanding and appreciation related to our audience's likely starting points and experiment with new ways to envision the temporal and spatial dimensions of science's stories. We can learn from engaging stories that have a micro-ring scope as potential options for adapting our storytelling techniques to foster an appreciation for both small and large scopes. We can learn from meso-ring and macro-ring sequence stories how to creatively adapt the ordering of events and information. Because each story will have different audiences, topics, and contexts, time and space can be adapted to best suit the goals of the storyteller.

Some science storytellers may balk at disrupting sequences of time or purposefully adjusting a scientific controversy's frame of reference. However, overly restricting the rings we use for sequence and scope may also lead to issues with comprehension, perceptibility, and feelings of certainty. Therefore, it may be sometimes appropriate, depending on audience, topic, and context, to adjust temporal and spatial considerations to incorporate various rings. Using the sequence and scope wedges as hub spirals may be useful for scientific stories seeking to foster a sense of immediacy, urgency, and relevance between audiences and the story or in instances where specific examples are plentiful and in line with the goals of the storyteller.

4 Storyteller and Content Wedges

"Climate Barbie," "ugly fake scientist," "old thin-looking hag," "stupider than a cow"—these were comments levied against women climate scientists and communicators such as Georgia Tech professor Kim Cobb and Texas Tech professor Katharine Hayhoe.[1] Combining insults of intelligence with insults of appearance, climate skeptics personally attack those who speak about climate change, especially women. Advocating for climate change as a public figure positions one in the crosshairs of climate skeptical attacks on one's character, credentials, and personhood. These terms specifically target women as incompetent and unintelligent, which makes them unacceptable contributors to the climate change discussion.

When my first book was published in 2019, I quickly became familiar with "global hating," or the insults and online abuse levied against those who publicly support climate change mitigation.[2] After interviews or pieces of media coverage related to my book, I often received a flurry of messages from upset and sometimes enraged people, predominantly climate skeptics, who strongly disapproved of my work. I was told that I supported climate change to "force" my personal opinion onto others and that I was complicit in "repressing/crushing dissent" on climate change. On an article that I wrote for *The Conversation,* one commenter told me to "shut your damn mouth up" because "you haven't done any research or have any critical thinking" and concluded their comment by calling me a "DUMBASS." In these emails and comments, my credentials and intentions are often questioned, with one person asking me, "What gives you the right to comment on [climate change]?"

From being accused of being paid off by the Environmental Protection Agency to being yelled at in ALL CAPS to being told to kill myself, I have learned firsthand how aggressive and even threatening people can be toward climate change advocates, especially when hiding behind a keyboard. Public scientists are frequent targets of harassment, especially in digital spaces.[3] Communication scholar James Zappen referred to the "hostile expression of strong emotions" online as "flaming."[4] My colleague Denise Tillery and I located flaming as a practice common in climate skeptical Facebook groups.[5] These attacks are present in other scientific controversies, too. In the COVID-19 pandemic, many medical and health scientists have been faced with threats of physical and sexual violence.[6] Even studies conducted prior to the start of COVID-19 have found that as many as one in six female physicians have been targets of online harassment.[7]

These behaviors are not only limited to online harassment and threats.[8] Physician Krutika Kuppalli, for example, received a phone call from someone threatening to kill her after she testified to a US congressional committee. Belgian virologist Marc Van Ranst and his family were moved to a safe house after receiving a note that they were being targeted, and UK chief medical adviser Chris Whitty was assaulted and put in a headlock while walking through a London park. Public medical figures in Mexico, India, Myanmar, and many other countries have been subject to vicious attacks, such as having hot coffee and bleach thrown on them and having their COVID-19 wards vandalized.[9]

Why does expressing scientific information draw forth such heightened emotions, physical violence, and sexist and racist attacks? Although we may try to limit the content of science communication to purely technical topics, science communication necessarily intersects public concerns and a wide variety of political, economic, religious, and cultural beliefs.[10] Tillery and I have argued that scientific topics such as climate change are discussed frequently in concert with political issues such as gun rights, immigration, LGBTQ+ rights, and feminism.[11] To ignore these potential implications of scientific messages can be isolating for audiences and provide fodder for skeptics to claim that scientists are removed from the real world. Content and conclusions that go against someone's worldview can be ascribed to the storyteller, thereby directing anger, hate, and even violence onto individuals for their perceived relationship to a topic.

The storyteller and the story they tell are inextricably connected. Education scholar Margaret Kovach (Plains Cree and Saulteaux) wrote that "stories can never be decontextualised [*sic*] from the teller."[12] Stories are sourced from the storyteller, who actively shapes and expresses the content to communicate knowledge and construct relationships.[13] Consequently, dis/trust of particular storytellers can cast a positive or negative light on the information in the stories they tell. For example, political loyalties or religious faith could invite belief in a story regardless of the story's content.[14] In analyzing the rhetoric of former US president Donald Trump, communication scholar Bonnie Dow argued, "Voters were attracted to who he was and how he presented himself as much as or more than what he said," an example of how the storyteller influenced acceptance of their content.[15] It is important to note that one's credibility as a storyteller is not fixed. One may be qualified to tell a story on a particular topic related to an area of expertise or experience but be unqualified (or viewed as such) by an audience when talking about a different topic.

Many people and groups can tell science's stories. Scientists are prominent storytellers, but so are science journalists, political figures, social media users, members of the public, advocacy groups, and others both within and outside the formal bounds of scientific institutions. Some of science's stories are *about* storytellers, meaning that science's storytellers may also be characters in the narratives of others. For example, Dr. Anthony Fauci was a prominent storyteller during the COVID-19 pandemic in the United States under Trump's presidency. People's faith in Fauci as a storyteller, from national hero to calls to #FireFauci, was in part influenced by stories about him from other storytellers. In the short film *Plandemic,* Judy Mikovits falsely accused Fauci of having financial stakes in COVID-19 vaccines and thereby of bias in his role

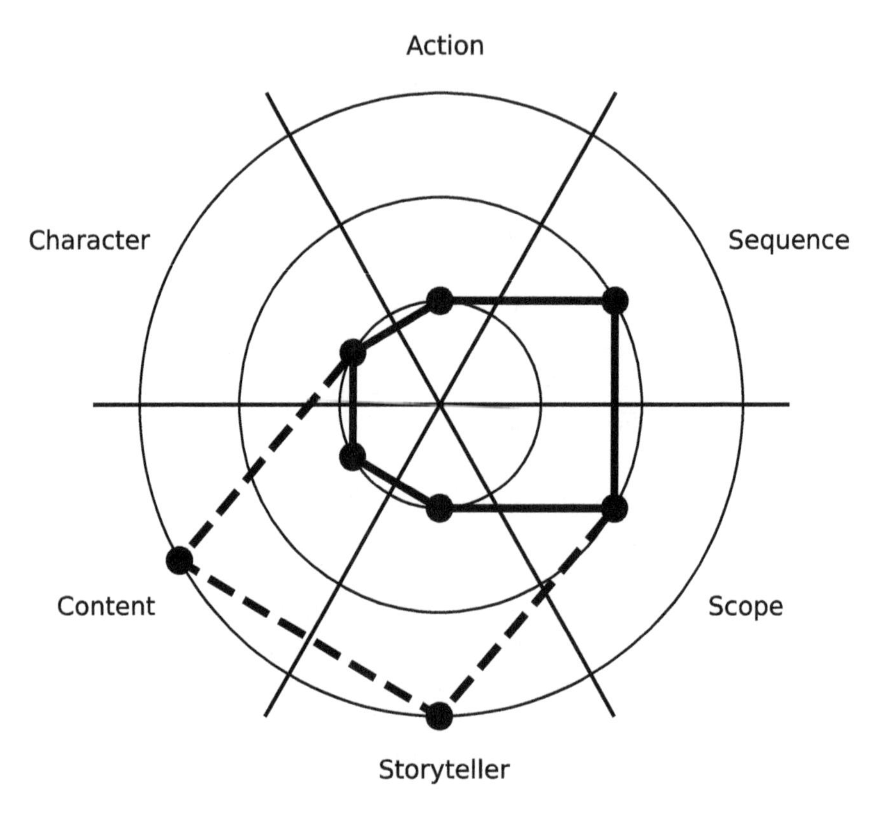

Figure 6. A narrative web of Dr. Anthony Fauci's final White House press briefing about COVID-19. The constellation for the audience of Fauci supporters is mapped in a solid line, and the constellation for the audience of Fauci skeptics is mapped in a dashed line, showing different interpretations of his trustworthiness and credibility.

as chief medical advisor to the president. Although repeatedly debunked by many online sources as inaccurate, false, and perpetuating lies about vaccines, *Plandemic* garnered more than eight million views and emboldened anti-maskers and anti-lockdown protesters.[16] These stories, among others, contributed to skepticism of Fauci as a character in COVID-19 hesitant stories, which then carried over to his role as a storyteller.

In Fauci's final press briefing in his role as chief medical advisor, he tells a story of urgency and encourages everyone, no matter their ideological differences or what political party they are a part of, to get vaccinated against COVID-19. Fauci emphasizes individuals' safety in protecting themselves from the virus and invites audiences nationwide to take clear preventive action for future health and well-being. For some, Fauci's credibility is specific and trustworthy, and the topics of health and safety are relevant values to their everyday lives. For other audiences, Fauci is not trustworthy and instead represents the abstract and nebulous workings of conspiracy and unscientific self-interest, casting doubt on the relevance and accuracy of his story. Accounting for different audience responses to Fauci as a storyteller, the narrative web for his final White House press briefing (see figure 6) reflects different mappings of storyteller and content. Narrative webs are always mapped in relation to an audience and context to reveal important patterns and aberrations.

In this chapter, I focus on rival stories that characterize science's storytellers, and thus their content, as corrupt and untrustworthy. Like all the wedge pairs, the storyteller and content wedges influence one another, but I have separated them to discuss the unique features of each. The chapter discusses three themes related to the storyteller wedge and one theme related to the content wedge. The three storyteller themes are (1) scientific dishonesty, (2) financial gain, and (3) political agendas. The content theme is irrelevant and incomplete content. I provide examples across the four case studies to examine how these themes influence the storyteller and content of scientific stories when faced with disingenuous rival stories that seek to undermine them.

THE STORYTELLER WEDGE

The storyteller wedge is related to the trust an audience has in the person telling the story and their credibility. For example, some stories may be told by individuals, such as a parent telling a bedtime story to their child, while other stories may be historical narratives about the founding of a nation, or social narratives that produce stereotypes about marginalized groups. As mentioned previously, a storyteller may be within their own story, as in the case of an autobiography or the recounting of a memory. In that case, the storyteller and at least one of the story's characters may be the same. The same person/group, thus, can be evaluated based on their appearance as a character within the story and the credibility they give to the story as its storyteller.

Walter Fisher refers to evaluations of the storyteller as a story's "characteriological coherence," which relates to the audience's "regard for the intelligence, integrity, and goodwill (ethos) of the author, the values [they embody] and would advance in the world."[17] A story's characteriological coherence is related to ethos, or how a storyteller appeals to their credibility or trustworthiness to speak on a topic. In other words, characteriological coherence is measured by "the degree to which the narrator is taken to be a character whose word warrants attendance."[18] Fisher argues that characteriological coherence is communicated through three values: consistency, completeness, and character.[19] A storyteller must be consistent, not contradict themselves, and be a reliable source of information. A storyteller must have complete knowledge and be able to speak to the full story. Lastly, a storyteller must be of strong character, with good reasons and values.

Using qualities similar to Fisher's, rhetoricians Pamela Pietrucci and Leah Ceccarelli analyzed the ethos of scientists as garnering trust if they embody moral values, goodwill, and practical judgment.[20] As discussed in chapter 2, Robert Merton's norms of scientific inquiry, which we ascribed to scientific characters, may also be ascribed to the storyteller. Instead of a character within a story, the storyteller may be accused of failing to meet a standard of scientific inquiry, thereby raising questions about the true "scientific" nature of the information and casting doubt on the content therein.

If the storyteller is named and is trusted, reliable, and an authority on the story's content, the storyteller wedge can be mapped on the micro-ring. Even if a storyteller is known and named, a story may not have characteriological coherence if that particular storyteller is not trusted. For example, Fauci and the Centers for Disease Control may appear on the surface to be micro-ring and meso-ring narrators, respectively, of COVID-19 stories, but both garnered distrust from some portions of the US population, which may result in a macro-ring placement for certain audiences.

An unnamed, implied, or obscured storyteller could contribute to a lack of narrative rationality as audiences are unable to assess the storyteller's reliability, resulting in a macro-ring placement. An abstract storyteller may fail to connect with audiences or may be distrusted due to associations with that entity. When trust is put into question or has been lost, the wedge may be mapped on the meso-ring or macro-ring, respectively. Distrusting or being skeptical of a storyteller will naturally affect an audience's willingness to adopt the story and accept it as a meaning-making resource.

My analysis of science's storytellers is informed by the "corrupted scientist archetype," which rhetorician Doug Cloud identified in congressional hearings on climate change.[21] The original conceptualization of the archetype includes money and politics as two types of corrupting influences that undermine the character of climate scientists, and, consequently, climate science. The trope of "corruption" shifts conversation away from the science and onto motives, or *why* science communicators are telling the story and what they are perceived to gain by doing so. If people believe these assertions, or at least listen enough to feel skeptical, the storyteller's credibility is undermined and the content of the story may be questioned as being false, exaggerated, or harmful to its audience. While some people will trust science storytellers and dismiss these claims of corruption, others may find accusations about scientists' motivations convincing, at least enough to create hesitation in adopting science narratives' content.

I identified three themes in how science's storytellers are undermined by rival stories: (1) scientific dishonesty, (2) financial gain, and (3) political agendas. The first theme presents science storytellers as wrongfully excluding minority viewpoints and being indoctrinated by the current system to uncritically accept information as given by the establishment. These storytellers are not upholding the standards of science and thereby are violating the norm of "organized skepticism" in scientific inquiry.[22] The second theme examines accusations that science's storytellers financially benefit from their work and are thus influenced to conduct research a certain way or reach particular results. The third theme is related to political agendas and claims that science's storytellers are not influenced by data and evidence but by political loyalties. Themes of financial gain and political agendas undermine the value of "disinterestedness" in scientific inquiry.[23] All three types of claims discredit science's storytellers and thereby taint the quality, validity, and reliability of the content within the story.

THE CONTENT WEDGE

While the scope wedge attends to the breadth of temporal and spatial frames, the content wedge attends to the breadth of topics and information within the story. The content wedge is also related to the narrative fidelity of the story and whether its topics, morals, and general components resonate with the experiences of the audience. Crafting content that is relevant to audiences may mean extending outside technical realms and considering how other arenas, such as people's daily lives, health, safety, education, jobs, and well-being are affected by scientific content. For some topics, like COVID-19 and climate change, these links may be more evident than others. That does not mean, however, that science communicators are incorporating that information in the stories they tell. A perceived separation between technical and public issues, which philosopher Bruno Latour called the science-policy divide, can lead some scientists and science communicators to see themselves as removed from everything but scientific inquiry.[24]

Even if science communicators do not see certain issues as relevant to the performance of science, public audiences often interpret technical information through or at least alongside those priorities. Fisher called coherence related to content a story's "material coherence." Importantly, material coherence is not judged in and of itself; material coherence is "determined by comparing and contrasting a present story with stories told in other relevant discourses."[25] Important to material coherence is "completeness," because in comparing stories, audiences look for "factual errors, omission of important arguments, and . . . distortions."[26] Depending on which stories audiences already believe and the circulation of other stories, audiences are making many comparisons through which they judge the believability and applicability of science's stories. Fisher notes that there "is no story that is not embedded in other stories," so science communicators ignore rival stories, disingenuous and productive, to their own detriment.[27]

Stories that cover relevant and meaningful content can be mapped on the micro-ring, while less relevant and more abstract content can be mapped on the meso-ring and macro-ring. It is important to remember that content mapping, like the other wedges, will always be relative. For example, Kenneth Burke argued,

> Even if one speaks very clearly and simply on a subject of great moment to himself, for instance, one is hardly communicating in the desired sense if his auditor does not care in the least what he is saying. A philosopher, if he has a toothache, is more likely to be interested in dentistry than in mathematical symbolism.[28]

Things that are urgent (or made to feel so) will naturally capture increased attention. The ebbs and flows of the news cycle, environmental changes, social dynamics, and other events will influence what people focus on and what they view as meaningful. During the height of the COVID-19 crisis, some people felt the virus intimately, having contracted it themselves or having known people who had. Alternatively, those with little direct contact with the virus or who were asymptomatic carriers were less likely to express concern and subsequently follow health and safety guidelines.[29] Additionally, the emergence of COVID-19 dominated public thought and news coverage, subsuming other concerns such as climate change as less important and urgent.

Even if information is important, it may not easily be communicated in a way that makes it seem relevant to audiences. I conceptualize relevance through linguist Jeanne Fahnestock's concepts of "application" and "wonder."[30] Application refers to how information is linked to end results that directly benefit the audience, and wonder refers to how information may be praised or celebrated because it is linked to something already valued by the audience. A story about a new scientific discovery, for example, would use an application appeal to discuss how the discovery will improve people's daily lives and a wonder appeal to praise it as something "'never before' seen."[31] Based on what people prioritize, certain stories will be more engaging and relevant and discuss issues of concern to a population, whereas others will fail to capture their attention.

The fourth theme I discuss in this chapter is related to a story's content and material coherence. This theme, called irrelevant and incomplete content, focuses on claims that the content of science's stories is not relevant to people's lives. Furthermore, scientific stories may fail to cover important areas of life, such as religion, ethics, and the economy, and therefore may be seen as offering incomplete explanations on which to base decisions. These characterizations often come from disingenuous rival storytellers who propose that their particular story supplies a missing piece or is less incomplete than science's stories.

SCIENTIFIC DISHONESTY

A prominent claim across all four case studies is that science storytellers silence dissent and do not consider alternative viewpoints, thereby excluding some information from decision-making. Leah Ceccarelli argues that "appeals to open-mindedness, freedom of inquiry, and fairness create discursive traps" for science storytellers, because they posit particular values that science *should* embody, but which are unreasonably applied to scientific topics on which there is consensus, such as evolution and climate change.[32] Despite the unreasonableness of these claims, they can still be highly persuasive to audiences who are sympathetic to the silenced voices or are uncertain about which storyteller is more trustworthy.[33] Ceccarelli argued that it is a tactic of science skeptics to "describe academic practices like peer review and tenure as mechanisms for an orthodoxy to suppress those who have a dissenting view."[34] Claims of gatekeeping and exclusionary practices are particularly harmful to the credibility of science storytellers because the claims challenge whether scientists are unfairly dismissing alternative information. It is decidedly unscientific to stop questioning and searching for more knowledge, so by framing science storytellers as closed to questions or modifications, skeptical storytellers can cast doubt on science's characteriological coherence.

Climate skeptical groups note that scientists who embrace a starting point that is not anthropogenic (human-caused) climate change are not published as frequently in mainstream journals. For example, Calvin Beisner, president of the Cornwall Alliance, a religiously based climate skeptical organization, argued that research coming from mainstream scientific journals should be suspect. He noted, "Peer review has been irretrievably compromised" because "a scientific orthodoxy took over."[35] The peer review process has been tainted to the point that "dissenting scientists . . . were cowed into silence" by the predominant, naturalistic worldview.[36]

Joseph Bast, president of the Heartland Institute, also argues that climate skeptics are purposefully kept out of academic publishing. He notes that "the average [climate] skeptic has been published about half as frequently as the average alarmist." Bast does not attribute the publication discrepancy to quality or "credibility," but to "publication bias" toward articles that support causation, as opposed to ones that falsify hypotheses, and editorial bias from within the community to not publish climate skeptical research.[37] Climate skeptical groups frame scientists who deny the consequences of climate change as a persecuted minority who are excluded from academia due to bias and discrimination, which thereby shapes the contours of appropriate and publishable climate science.

Academic institutions are also identified by creationist groups as exclusionary spaces that purposefully exclude those who would go against the scientific norm. Creationist group Answers in Genesis argues that the persecution of Christian thought happens frequently in academia, which silences creationist perspectives. In summarizing a poll of creationist academics, Answers in Genesis contributor Jerry Bergman argued, "Almost all felt that they had faced serious religious discrimination in their academic careers at least once or more often" and "all, without exception, felt that openly holding a 'scientific creation' worldview would seriously impede or terminate an academic career."[38]

Creationists argue that "ruling out dissenting voices on the evolution question, even religious ones, amounts to viewpoint discrimination," and instead all views, especially creationism, should be included in the debate.[39] The Institute for Creation Research makes frequent reference to science as an open, academic community that should always be willing to listen to the opinions of other scientists. The Institute for Creation Research describes itself as encouraging "scholarship, investigation, and careful scrutiny of origins concepts."[40] Institute for Creation Research contributor Christine Dao argued, "Creation science researchers are willing to examine the data and, if necessary, move on to more interesting and securely justifiable discoveries."[41] Thus, the Institute for Creation Research presents itself as better embodying scientific principles of open-mindedness when the data do not support prior conclusions. Institute for Creation Research president Henry Morris accused evolutionists of being "adamantly committed to anti-creationism," closed to debate, and reluctant to hearing alternative explanations.[42]

An implication of restrictive educational systems is that individual scientists are brainwashed to follow those systems without questioning.

Discourses from vaccination and COVID-19 skeptics often portray science storytellers as indoctrinated to follow what they have been trained to do medically. Instead of focusing on structures of peer review and academic hiring, claims about vaccination and COVID-19 storytellers center on a perceived lack of knowledge, incorrect medical training, and a failure to consider alternative medication as violations of scientific integrity. Many medical organizations acknowledge the problem of credibility. For example, the American Academy of Pediatrics argued that pediatricians face "increasing challenges in conducting" their work, including "vaccine hesitancy."[43] In an article published in *American Family Physician*, medical researchers Jeanne Spencer and colleagues noted that "some parents express concern that physicians are not well educated on the adverse effects of vaccines or that physicians purposefully withhold information on adverse effects."[44] Contrary to norms of viewing doctors as reliable authorities, some parents turn to "websites containing inaccurate information" as more reliable sources.[45]

Many vaccine skeptical websites argue that doctors are indoctrinated to parrot the safety of vaccines, and thus do not question them. Similar to the silencing of alternative scientific views seen in the evolution and climate change controversies, vaccine skeptics claim that "doctors who question vaccines are publicly humiliated or scorned," receive "ridicule and criticism," and even risk having their licenses "revoked."[46] Vaccine skeptics claim that doctors are influenced by "negative peer pressure" to advocate for vaccines even when they disagree with them.[47]

Citing a list of credentialed doctors who speak out against vaccines, VacTruth argues that "even though they attended eight or more years of medical school," doctors "often lack adequate education about vaccines and the information that is provided [to them] is often funded by pharmaceutical companies."[48] VacTruth's article cites a vaccine skeptic who has a medical degree, Suzanne Humphries, as saying, "[In medical school] we learn that vaccines need to be given on schedule. We are indoctrinated with the mantra that 'vaccines are safe and effective.'"[49] Based on these examples, VacTruth concludes that "most pediatricians have never read a vaccine package insert and don't know how to recognize an adverse reaction to vaccines."[50] Furthermore, doctors "fail to do their own research" and uncritically follow Centers for Disease Control guidelines.[51]

A different but similarly named vaccine hesitant group, VaxTruth, argues that the government "protect[s] the vaccine-manufacturers, who have been granted complete immunity from liability if their vaccines happen to kill or permanently injure your child."[52] VaxTruth repeats this claim of legal immunity across its website: "Vaccine manufacturers are not liable for any harm their products cause. They're protected by the U.S. government. They have no incentive to make safer products."[53] VacTruth also includes doctors in its statement of liability, noting, "Vaccine manufacturers (and doctors!) are given complete immunity from any legal liability if your child is harmed by their vaccine(s). No other industry enjoys this level of protection from a product that could injure a child!"[54] Discussions of liability, or lack thereof, signal to audiences that the medical community can advocate for potentially harmful treatments with no fear of repercussions.[55] These claims frame doctors and pharmaceutical companies as unmotivated to act in a patient's best interests, thereby violating medical oaths and scientific integrity.

The ongoing controversy over COVID-19 also includes the silencing of certain voices and claims of censorship. In a series of interviews with frustrated medical professionals, an article in the *Epoch Times* discussed the "top down" approach that hospitals were following from the Centers for Disease Control. Instead of doctors being able to make their own decisions, medical professionals were quoted as saying, "Something came down from on-high that only allowed us to do prescribed protocols. . . . If nurses or doctors stepped outside the protocols" there would be consequences, such as being "threatened or fired."[56] This article portrays doctors and medical professionals as being duped by the system and restricted from acting outside of the scope of official protocols. Later in the article, a medical professional noted doctors requested to shift protocols and try alternative treatments, but those "queries . . . consistently fell on deaf ears."[57] Even when doctors tried to adjust the prescribed protocols, they were unable to modify the system. Disingenuous rival stories to science's COVID-19 stories asserted that doctors were not only restricted in their medical protocols, but they were also prevented from advocating for alternative treatments or questioning the effectiveness of vaccine safety.

Unlike the other case studies, COVID-19 is a novel controversy only a few years in the making. Addressing the novelty of COVID-19, while accurate, may inadvertently undercut audience's faith in science and characterize decisions as based on nonscientific information. In one article, a doctor was quoted as saying, "You have to remember, COVID was new to us. . . . These are things we'd never done before."[58] If a virus is novel, then naturally the advice for how to treat and respond to the disease will also change. This is one of the hallmarks of science; it is flexible, dynamic, and constantly updating with new information. When audiences look to science for all the answers, however, storytellers can be caught in a discursive trap of seeming to be capricious and vacillating if the medical advice changes. A search of "flip-flop Fauci," for example, leads to hundreds of thousands of hits on Google, with headlines such as "Flip-flop Fauci: The purveyor of doom and gloom," "Did Fauci flip flop or follow the science?," and "Fauci Flip Flops on Lab-Leak Theory."[59] Claims of flip-flopping can undermine the credibility of scientists by framing them as misinformed or as making decisions on whims rather than evidence and scientific integrity.

Disingenuous rival stories across all four controversies accused science's storytellers of being inappropriately exclusionary toward alternative views, which compromises credibility. Disingenuous rival storytellers framed removing certain voices from scientific deliberations as inherently anti-science and against freedom of inquiry. In a paper I coauthored with Tillery, we called this practice appealing to "hyperrationality," which is when skeptical groups undermine the "rational standards" of scientific practice by substituting a different "resource that is more rational than mainstream scientific argument."[60] In other words, hyperrationality occurs when science as a practice is framed as irrational and skepticism is thereby framed as the more rational and logical perspective. If mainstream science is seen as unscientific or as failing to follow its own norms and standards by silencing others, then disingenuous rival storytellers can step in to fill those gaps for public audiences.

FINANCIAL GAIN

Cloud described accusations of financial corruption as the claim that "scientists do bad science for money."[61] The trope of monetary corruption is quite convincing because it provides a compelling motive behind the telling of science's stories. Tillery identified the trope of "follow the money" as one of the most "damning critiques of science" in climate discourse."[62] For example, a climate skeptical op-ed reads, "It was never about climate change—it was about powerful and moneyed elites controlling our economy and setting our lifestyles back to the Stone Age under the guise of preserving the planet from human destruction."[63] Michael Shellenberger, author of *Apocalypse Never,* argues that environmental groups hypocritically accept "hundreds of millions of dollars from fossil fuel interests,"[64] attributing climate scientists with financial motivations and failing to follow their professed environmental values. Collecting money becomes a reason why climate scientists do science and why they might do science a particular way.

An article published by the Heritage Foundation titled "Follow the (Climate Change) Money" argued that "one reason so many hundreds of scientists are persuaded that the sky is falling is that they are paid handsomely to do so."[65] The article continued by arguing that there has been a "tidal wave of funding" for climate change programs and research that "reveal[s] a powerful financial motive for scientists to conclude that the apocalypse is upon us."[66] An article in the *Christian Science Monitor* by Robert Rapier also commented on the influential force of money, noting, "I do question how money influences some of the environmental organizations" and "their objectivity."[67] The simple suggestion of a correlation between money and scientific conclusions can cast doubt on a storyteller's credibility and thus their information.

Heartland Institute president Bast expressed concern over grant funding as influencing climate science:

> In the case of climate change, hundreds of millions of dollars in government grants have gone to scholars who say they are trying to find a discernible human impact on climate. . . . Much less funding is available to scholars who say they are seeking to find natural causes for climate change, or explanations of natural phenomena that don't involve climate change.[68]

Bast's argument is not only that people who perform climate science research are well funded, but also that scientists who do not support climate change are not getting funding, implying that scientists may adjust their research questions to access funds. The distribution of funding is thus linked to gatekeeping within academia; only certain viewpoints are funded and published, further silencing dissent and motivating adherence to scientific narratives of climate change through financial motivation.

Public environmental figures are also accused of acting for monetary gain. For example, former US vice president Al Gore is a frequent target of climate skeptical groups for his perceived financial gain from climate activism. One article noted that Gore's net worth is over $300 million "largely due to his climate change advocacy and his embrace of the alarmist position on global warming,"[69] directly attributing Gore's wealth to the position he takes on climate change. Pointing out the size and extravagance of Gore's home, for example, is weighed against Gore's claims of a "carbon-free lifestyle."[70] In this narrative, not only is Gore profiting from telling climate change stories, but he is also requesting the public make sacrifices he is unwilling to make himself. Gore is just one example of climate activists who are accused of profiting from their advocacy and, therefore, are unduly influenced by financial gain as told in climate skeptical narratives.[71]

Vaccine skeptics propose financial motivation as a primary reason to doubt the stories told by medical professionals. VaxTruth argues, "Of course to you, your child is priceless—with value far beyond what money can buy. But many in these different groups [official medical bodies] see only dollar signs."[72] Using an analogy to compare administering vaccines to selling cars, VaxTruth notes,

> The CDC makes money off of the sale of vaccines. The CDC and vaccine makers are the ones who are educating your doctor about vaccines. It's like taking the word of a used car salesman. "Trust me! This 1976 Ford Pinto is perfectly safe!"[73]

In associating the Centers for Disease Control with used car salespeople, VaxTruth portrays the Centers for Disease Control as untrustworthy scammers charging for things that are unnecessary or even harmful. VacTruth summarizes this point clearly: "Indeed, the love of money (greed) is a very common human attribute" affecting even medical professionals who should be committed to promoting health.[74]

Profiting from vaccines was a claim that emerged in the COVID-19 controversy as well. In particular, public figures such as Fauci and US President Joe Biden have been accused of promoting vaccine uptake due to personal stakes in vaccine manufacturers. Fauci was accused of purposefully failing to recommend alternative treatments for COVID-19 because he has invested "100 million dollars" in a coronavirus vaccine and needed to create demand for the product.[75] Although these claims were easily debunked, it did not stop them showing up across COVID-19 skeptical social media platforms, including the short film *Plandemic,* where scientists including "Tony Fauci" were described as "stand[ing] to make hundreds of billions of dollars" from vaccine production.[76] In June 2022, Senator Rand Paul (R-KY) interrogated Fauci in a Senate hearing about where Fauci or other National Health Institute employees were getting royalties from vaccine development. This "sparring match" between Paul and Fauci was covered in many media outlets, which spread the possibility of a financial incentive even in denying that one existed.[77]

Members of the Biden administration have been accused of having "a financial stake in or professional ties to vaccine manufacturers" and opposing policies that would limit their profits.[78] In an article written for *The Intercept,* Lee Fang outlined ten perceived conflicts of interest, including Secretary of State Anthony Blinken's consulting work for Gilead Science, which manufactures remdesivir, a COVID-19 treatment medication.[79] The article notes that the "Biden administration is primed to favor profits over people," squarely positioning the stakes of financial gain against the risks to people's lives.[80]

In vaccine skeptical stories, hospitals stood to gain money by prescribing certain COVID-19 treatments. One article referred to the "financial incentives" that hospitals could get: up to $15,000 for every patient on remdesivir and up to $40,000 for patients on ventilators.[81] These numbers were contrasted with financial reimbursements for prescribing hydroxychloroquine and ivermectin, which were four dollars and one dollar, respectively.[82] Claims about government reimbursements for medical expenditures related to COVID-19 are rooted in some truth. Hospitals did receive money from emergency government funds based on how many patients were diagnosed with COVID-19 or were on ventilators because this information demonstrated a need for funding to purchase additional supplies, not to line the pockets of administrators and doctors.[83] However, this did not stop cherry-picked statements from doctors about COVID-19 funding taking on a life of their own on social media.[84] If people believe that hospitals have an incentive to put people on ventilators and diagnose them with COVID-19, then they may discount the severity of the illness or reports of the number of people who have caught the disease. Despite frequent fact checks that there was "no evidence of fraudulent reporting" by hospitals,[85] narratives of financial wrongdoing can be compelling because they ring true to our perceptions of greed's persuasiveness and pervasiveness.

Themes of financial motivations appeared most often in narratives of climate change, vaccines, and COVID-19. The rival stories of creationism and intelligent design did not tend to corrupt evolution's storytellers through financial appeals. More often used were claims of scientific and academic dishonesty, as previously discussed, and political agendas, to which we now turn.

POLITICAL AGENDAS

The trope of politics characterizes scientists as doing "bad science to 'toe the party line'" and support a particular political party's agenda.[86] Tillery refers to this trope as the claim that "science is ideologically based and enmeshed in politics."[87] The political form of the corrupted scientist trope often manifests as claims that science supports a liberal agenda and thereby adjusts scientific findings to support liberal policies, but not exclusively.

In the climate change controversy, the Acton Institute argued that environmental science is at its "best" when it is "dispassionate," but current climate science has become "a mouthpiece for [one's] political convictions."[88] A frequent "political mouthpiece" identified is the Intergovernmental Panel on Climate Change, which has been described as contributing to the "politicization of science," which undermines its credibility.[89] The Nongovernmental International Panel on Climate Change, a climate skeptical organization that spoofs the Intergovernmental Panel on Climate Change's name, argues that "there is ample evidence" that Intergovernmental Panel on Climate Change reports are "fueled by political considerations rather than actual science."[90] The organization is thus characterized not as a scientific organization, but as a politically motivated group that produces tainted science.

The Cornwall Alliance reports that environmental groups such as the Intergovernmental Panel on Climate Change conspire with "left-wing, pro-abortion, pro-population control" advocates.[91] The Heartland Institute described climate science as being run by "an increasingly activist, liberal media complex."[92] The Heartland Institute makes frequent reference to political leaders such as Gore, former US president Barack Obama, and John Holdren, a democratic science advisor, as advancing climate policies in order to achieve political ends.[93] Climate scapegoats such as the Intergovernmental Panel on Climate Change, Gore, and even Bill Nye, an engineer and public science figure, are leveraged as evidence that political loyalties, primarily or solely, motivate science's conclusions about a changing climate.[94]

The Heartland Institute argued that climate advocates are motivated to make themselves a class of "ruling elites" who "will be in charge" of everything.[95] Calling attention to scientific discourse as a story, this article accuses climate advocates of forwarding a "phony 'saving the planet' narrative" in order to achieve "total control over our lives, livelihoods, living standards and liberties."[96] The Cornwall Alliance shares this concern about control by noting that climate advocates strive for a "dangerous expansion of government control over private life,"[97] an appeal to "big government" policies associated with liberal politics.

The "politics" of the human origins controversy is rooted in the tension between the ideologies of "materialism" and "supernaturalism," Answers in Genesis contributor Roger Patterson notes.[98] The materialist side is intertwined with atheism, secular humanism, and liberalism, which Answers in Genesis groups together as "left" orientations against more conservative and Christian views, especially on social issues such as abortion.[99] Answers in Genesis sees itself and other Christian organizations as under attack by "liberals and liberal organizations" who "spew hateful, demeaning, profanity-laced language" at creationists.[100]

Creationist voices frame evolutionary scientists as wanting to destroy religion and thereby being influenced by their loyalty to a materialist worldview and the "religion of atheism."[101] In an article about a Smithsonian exhibit about human origins, Answers in Genesis CEO Ken Ham wrote that the exhibit's purpose was "not only to indoctrinate children and adults in evolution, but also atheism."[102] Ham notes that the exhibit forwards "explanations according to their view of naturalism, the religion of atheism."[103] In another article, Answers in Genesis writer John Altevogt argued that "information provided by liberals" is "deliberately misleading" regarding evolutionary science.[104] Additionally, Altevogt accuses liberals of having a "political agenda" that causes them to overlook inconsistencies in evolutionary science, indicating that it is politics instead of scientific inquiry guiding belief in evolution.[105]

For climate change and evolution, there tend to be clear conservative versus liberal lines drawn. However, the controversy over vaccination, prior to COVID-19, had not typically been subject to political polarization. In 2015, the Pew Research Center found similar beliefs on the safety of vaccines across self-identified Republicans, Democrats, and Independents.[106] Indeed, the more strongly someone was affiliated with either conservative or liberal politics, the "more likely [they] were to believe that vaccination is unsafe."[107] People who oppose vaccination come from both sides of the aisle for different reasons. Conservative vaccine skeptics tend to oppose government overreach and may turn to religious sources of healing instead of vaccines. Liberal vaccine skeptics tend to oppose what they view as unnecessary medical interventions and may use alternative medicines instead.

The lack of political polarization around vaccines changed drastically during the COVID-19 pandemic, where reactions to vaccination and other COVID-19 preventative measures such as stay-at-home orders and mask mandates were largely split down political lines. When medical professionals provided guidelines on prevention and treatment measures, they were implemented and communicated to constituents at state and city levels. Politicians became science storytellers who translated medical advice to their constituents to justify their COVID-19 policies, or lack thereof. A study by communication scholars P. Sol Hart, Sedona Chinn, and Stuart Soroka found that politicians appeared much more frequently than scientists in newspaper coverage of COVID-19, which shifts COVID-19's storytellers into an undoubtedly political arena.[108]

Part of the political polarization of COVID-19 has been attributed to Trump, who, in the beginning of the pandemic, downplayed the risks of the virus by calling it the Democrats' "new hoax."[109] These statements framed the pandemic from the beginning as political fodder and emboldened Trump's followers to disregard the warnings and medical advice of his own health officials. Trump often refused to wear a mask; the flaunting of his maskless face was a form of what Nicholas Paliewicz and I call "indecent public exposure," which puts others at risk driven by a desire to display tough, rugged masculinity.[110] Following Trump's lead, Republican congresspeople, such as Representatives Marjorie Taylor Greene and Brian Mast, were repeatedly fined for refusing to wear masks. In total, unmasked congresspeople racked up tens of thousands of dollars in fines while the mask policy was in place.[111]

There were also political tensions regarding states' reopening. Many Republican governors pushed to reopen their economies after 2020 stay-at-home orders and filed lawsuits against Biden's federal vaccine mandate.[112] Tensions between federal and state control over COVID-19 protocols were also felt within states. With hesitation from leadership to create federal guidelines, policies were implemented in patchwork across the United States. Democratic mayors of large cities pushed back against Republican governors for reopening too quickly, such as the case of Atlanta and Columbia in Georgia.[113] Many Republican governors were accused of reopening states too early, contrary to what data suggested was prudent. Politics thus seeped into decision-making at a variety of levels and complicated medical advice on the best ways to mitigate the spread of COVID-19.

Political themes also spanned international politics. National polarization between Democrats and Republicans was replaced with China versus United States sentiments. Trump infamously referred to COVID-19 as the "Chinese virus" and the "kung flu," which led to upticks in anti-Asian sentiments.[114] Conspiracy narratives of the virus originating in a Chinese lab and being "leaked" intentionally or unintentionally also contributed to rising tensions with China. In response, conspiracies of the virus originating in a lab in North Carolina flipped the script and heightened fears of COVID-19 as a US-based bioweapon.[115] These narratives were circulated in the United States through "right-wing news organizations" and politicians such as Senator Paul.[116] Although the current explanation of COVID-19 is that it jumped from a bat to a human in the Huanan Seafood Market, lab-origin narratives gripped the world in the early months of the pandemic as people struggled to explain the virus and its origins.

The topic of political agendas was a prominent theme in the rival stories to climate change, evolution, and COVID-19, which undermines the credibility of the storyteller by associating them with political loyalties that cloud an "unbiased" interpretation of the science. When a scientific storyteller is corrupted through claims of dishonesty, greed, or bias, the content of their story is also compromised. The next section focuses on claims unique to content that scientific stories leave out important topics or fail to address top concerns of public audiences.

IRRELEVANT AND INCOMPLETE CONTENT

Because science's stories often emerge from technical spaces, their content tends to be rich with information, facts, statistics, data, and calculations. These scientific controversies affect people, some in more direct, urgent ways than others, but science's stories may not emphasize these connections in favor of more formal, instructional communication. As previously discussed regarding the sequence and scope wedges, the is/ought boundary is a perceived line of what is appropriate for scientists to discuss. The is/ought boundary involves measures of uncertainty regarding proposed actions and the future, but it also relates to appropriate content; scientists should discuss facts and information but not advise on policy. Audience expectations of what science's stories are supposed to cover and what is inappropriate to address can influence credibility when science's content oversteps these perceived boundaries.

Climate change's stories, as one would expect, center scientific predictions and long-term weather and temperature patterns. For many people lucky enough not to be under direct threat from natural disasters, these issues may seem intangible and irrelevant in one's daily life.[117] Indeed, even locations plagued by hurricanes, wildfires, and increasingly harsh and long summers may view these occurrences as aberrations and thus not indicative of long-term changes. People tend to view other topics as more important than the environment, including political, economic, and social issues, and separate climate change from those concerns. Instead of focusing on disciplinary conventions, Dylan Harris suggests that climate scientists can tell better stories by being "better listeners," thereby understanding the values and priorities of audiences.[118] Rival stories often "resonate" more strongly with the content audiences care about because those storytellers, Harris argues, "listen deeply to their communities."[119]

Disingenuous rival stories to climate change portray environmental science as lacking important content areas and values such as religious influence and economic input that their stories more fully address. Christian fundamentalist discourse promotes God's superiority over human actions, meaning that only God is powerful enough to harm or save the planet. For example, the Cornwall Alliance argues that the "Earth and its ecosystems—created by God's intelligent design—are robust, resilient, self-regulating," so cannot be affected by the actions of humans.[120] Climate science thus ignores the role of God in managing the climate. Indeed, some Christians believe that the Earth becoming warmer is part of God's plan, so do not see climate change as concerning or needing to be mitigated.[121] Climate science also inverts humanity's rightful place in a human-nature hierarchy. The Cornwall Alliance categorizes "subduing and ruling the Earth" as part of giving "glory to God."[122] To oppose this hierarchy is to transform humans "from ruler to slave of the Earth."[123] As a rival narrative, fundamentalist Evangelicalism may resonate with people who already ascribe to the Christian narrative and those who see faith as missing from climate science's stories.

Climate science is also accused of leaving out economics in advocating for mitigation policies. Think tanks such as the Heartland Institute and the Acton Institute propose alternatives to mainstream climate science by substituting scientific authorities with economic authorities. The Acton Institute views economics as a zero-sum game, arguing, "All of this [environmental advocacy] involves material costs, resources devoted to this advocacy that might also be used elsewhere."[124] The Acton Institute is concerned about environmental policies stopping economic growth, noting that the "anti-growth environmentalist agenda" will threaten economic empowerment of the poor.[125] The Heartland Institute lauds fossil fuels as the "foundation of economic growth and prosperity." Consequently, the group lobbies against taxing fossil fuels because doing so causes "economic growth to slow, makes food and other essential goods more expensive, and many of the good things we take for granted [would be] lost."[126] Economic rival stories center money as a value that resonates with many audiences. "Economic progress" has been described as "one of the many God terms of our age," meaning that opposition to it is often immediately dismissed as anti-American, anti-progress, and anti-business.[127] Stories that leave out economic concerns, therefore, may be missing an important value that audiences care about.

Evolution's storytellers are associated with having incomplete answers to the origins of life because they leave out religious explanations. In promoting such a view, evolutionary challengers have characterized evolution as immoral and anti-religious. Ken Ham, in his debate with Bill Nye at the Creation Museum, argued that evolution is "a naturalistic worldview that excludes the existence of God."[128] For evolutionary skeptics, the stories of evolution propose much more than a systematic explanation of life's development; it signals the "contamination" of public morals and children by "secular education, the secular media, and their secular friends."[129]

Creationists argue that the twin forces of evolutionary science and secularism are "intent on destroying Christianity," ascribing a harmful outcome of accepting the content of evolution's stories.[130] Some scientific organizations recognize these circulating accusations. The National Center for Science Education acknowledges that "creationists claim that evolution is intrinsically antireligious, a deadly threat to faith and morals."[131] Donald Prothero echoes that there exists "a general fear that science is a threat to religion," "that evolution is atheistic," and "that to even think about the evidence of evolution is sinful."[132] Answers in Genesis argued that belief in the evolution narrative would have drastic effects on human nature. Answers in Genesis president Ham accused evolution of supporting moral relativism, which he likened to the way that animals behave: "If we teach them [children] the whole universe is a result of natural processes and not designed by a creator God, they might be looking in the wrong places or have the wrong idea . . . that could totally influence the way they think."[133]

Disingenuous rival stories to evolution thus take issue with some of the content in evolution's stories, namely, their perceived advocacy of violence through the tenet of "survival of the fittest." Because there is an "odious myth" that "animals and humans are inherently selfish,"[134] some applications of Darwinian theory have extended selfishness and self-interest to nonbiological realms, including politics, society, and economics.[135] To adopt this "Darwinian view of life" is to see life as a "panorama of brutal struggle."[136] Thus, many "arguments against evolution" stem not from science but from "people who find it objectionable on religious or emotional or philosophical grounds."[137] Referencing social Darwinism, one scientific text noted that this philosophy "arose from a misguided effort to apply lessons from biological evolution to society," and that in the case of social Darwinism "science was misapplied to promote a social and political agenda."[138] The Pew Research Center described how "some evolution opponents argue that Darwin's ideas have proven socially and politically dangerous" because they "justify heinous crimes, from forced sterilizations to mass genocide."[139]

Some argue that a scientific view of origins removes morality from education and may lead to people treating fellow humans like animals. Scientific groups sometimes acknowledge these perceived links between evolution and immorality. For example, the National Research Council acknowledges the pervasiveness of these external viewpoints in noting, "From an early age, many young people in the United States absorb negative views about evolution" and "in many situations" may think that "evolution is evil."[140] Often citing Adolf Hitler and social Darwinism, some propose that the content of evolutionary science will exacerbate class tensions, create selfish humans, undermine values of mercy and societal unity, and reduce humankind to animals.[141] Because humans know themselves to be different from animals, evolution's stories may lack narrative fidelity in how they imply an entirely different picture of human nature and behavior.

Vaccine skeptical narratives accuse vaccination's stories of lacking key information, particularly by denying the role of parental influence. Instead of viewing credentials and experience as expertise, vaccine skeptics invoke different sources of credibility such as medical defectors and motherhood.[142] For example, VaxTruth argues that its website is more reliable than doctors, because "parents who are posting about vaccine dangers are trying to prevent other parents from making the same errors in judgment that we have made. We want to spare other babies the harm our own have endured."[143] This statement locates authority in the lived experienced of parenthood, driven by the noble purpose of protecting others. Piper-Terry often uses her position as a former medical researcher to complement her firsthand experience with what she perceives to be her daughter's vaccine injuries in order to juxtapose her credibility with that of doctors. On her website, VaxTruth, she argues,

> I was a researcher before vaccine-injury became the focus of my research. Just because I am a mom and just because my daughter was injured, that does not mean I lost my skills. Quite the opposite. It means my skills have been laser-focused on one issue for the last several years.[144]

Piper-Terry claims the authority that typically comes with being a part of the medical community but reframes her expertise as more trustworthy because she has left the medical community and struck out on her own as a mother with personal experience.

Parents who feel their child has suffered a vaccine injury may also find current medical explanations lacking or incomplete. The scientific stories regarding the relationship between vaccines and autism, for example, clearly demonstrate that there is no link based on immense amounts of scientific evidence.[145] However, such a narrative may fail to ring true for parents who have a child with an autism diagnosis and are looking for answers. Currently, scientific stories do not offer a clear-cut explanation for why some children develop autism and others do not. Shari Hoppin argues that many parents who advocate against vaccines are simply looking for answers regarding what causes autism.[146] She further notes that "for parents of autistic children, the lack of information about the cause of autism fuels a sense that medical experts and health authorities are, at worst, lying to them, or, at best, cannot be relied on to look out for their children's well-being" because they are incapable of discovering or revealing the full story of cause and effect.[147] Without a full explanation, science's stories about vaccines may ring hollow and fail to include the content that parents desire to know.

In the scientific dishonesty section, I discussed how people who advocated for alternative treatments to COVID-19 felt wrongfully excluded from medical practices. In this section, I want to more directly highlight the alternative medicines themselves as content missing from COVID-19's stories. Claims that medical science overlooked or did not consider certain medical options characterizes science's stories as incomplete or lacking the full content of available medical treatments for the virus. A few prominent names who advocated for alternative medications are Dr. Mehmet Oz and podcaster Joe Rogan. Oz was an early advocate of hydroxychloroquine and spoke about it on *Fox News* more than twenty-five times between March and April 2020 alone.[148] A statement from the campaign manager of Oz's Pennsylvania Senate run reads, "From the start, therapeutics meant to help with Covid-19 were regularly discounted by the medical establishment, and many great ideas were squashed and discredited."[149]

Rogan featured ivermectin on his podcast, *The Joe Rogan Experience*. With more than 200 million monthly downloads, Rogan's podcast regularly receives more airtime than mainstream news channels.[150] Claiming protection through free speech, Rogan downplayed the severity of COVID-19 by comparing it to the flu, promoted ivermectin, and claimed that young, healthy individuals are protected by their immune systems, so do not need to be vaccinated.[151] Oz's and Rogan's statements may have resonated with audiences who actively sought out alternatives to vaccines or simply had existing "suspicion of public health recommendations."[152] Excluding certain pieces of information and thus perceived to be failing to tell the full stories of COVID-19, science's stories were, in part, undermined by skeptical content.

In addition to claiming science's COVID-19 stories are incomplete, disingenuous rival stories also claimed that the stories left out important aspects of society, namely, personal freedoms and the economy. COVID-19 skeptics viewed a reliance on policies such as social distancing, lockdowns, and masking as inhibiting personal freedoms and incorrectly prioritizing public health over people's individual choices. COVID-19 skeptics saw lockdowns as "draconian" and "infring[ing] on their civil liberties."[153] Personal choice was weighted more heavily than public health by some, who saw their ideologies as a hallmark of the United States; one protester said, "[The United States is] not supposed to be the safest place on earth but it is supposed to be the freest." A frequent protest slogan was "Give me liberty or give me death," equating staying at home with sacrifices made during the American Revolution to fight for freedom from British rule.[154]

Opponents to COVID-19 measures associated mandates, particularly lockdowns, with harming the economy. Therefore, the economy was described as a topic not taken into consideration when COVID-19 measures were passed. One article discussed the record unemployment claims filed during lockdowns and pointed to interviews with protesters who said that "economic hardship [was] their primary reason for participating."[155] One protester who was interviewed said, "I'm gonna do what I got to do to feed my family. . . . If it means I got to risk my health then so be it . . . and yes even potentially the health of others." This sentiment was common during COVID-19 lockdowns, including the infamous statement by Texas Lieutenant Governor Dan Patrick that grandparents should sacrifice themselves for the US economy:

> No one reached out to me and said, "As a senior citizen, are you willing to take a chance on your survival in exchange for keeping the America that all America loves for your children and grandchildren?" . . . If that is the exchange, I'm all in.[156]

For some, the economy is the most important aspect of decision-making, even over people's lives. Those who find resonance with economic struggle may view science's stories of COVID-19 as ignoring that aspect of daily life.

Across all four controversies, disingenuous rival stories champion their ability to cover important topics such as economics, religion, and morality while making space for alternative explanations and solutions. Science's stories cover information in terms of its technical relevance, so may be missing the effects these topics have in other areas of life or the implications they have for audiences' values and priorities. Locating connections and overlaps between these topics can help science communicators bridge their content with what is most important to audiences.

CONCLUSION

The storyteller and content wedges reflect the source of the story and the main topics being discussed in it. A variety of people and groups can serve as scientific storytellers, and a variety of content is present in scientific stories, whether directly addressed or interpreted as relevant. Considering storyteller and content in tandem with one another can help science communicators reflect on the topics that go into their stories or should be added to those stories and who is the best source to communicate those stories. Without providing specific prescriptions for scientific stories, which will be situation, audience, and purpose dependent, I offer the following questions to brainstorm the storyteller and content wedges for science communicators and storytellers:

- Who is the source of the story? Who will be telling/communicating the story to the audience?
- Can I be the storyteller? What potential risks or opportunities does using myself as the storyteller open up? Am I willing to engage those risks by being a public spokesperson for a scientific topic?
- Is the storyteller similar to the target audience, trustworthy, willing to engage in conversation, and approachable?
- Is the storyteller well liked, trustworthy, and credible? What groups or individuals may be distrustful of the storyteller and why?
- Can the story be co-narrated or narrated by someone else to increase coherence and credibility?
- Is the content relevant, meaningful, and important to audiences? Do audiences care about the topics covered or are they made to feel as if the topics are personally relevant to them?
- Is my story leaving out important topics that my audience is concerned about?
- How can I make connections between the story's content and central values such as family, money, health, and safety?
- What rival stories does my story compete with? What stories circulate along with scientific stories?
- How do rival stories challenge, expand, or complicate the story I'm telling? How can I recognize these stories as collaborators and not competitors?
- How can we locate overlap between the interests and topics of science's stories and rival stories? Are there points of common ground or similarity at which these stories can converge?

These questions are meant to spark reflection and contemplation as opposed to prescribing an ideal web placement for the storyteller and content wedges. Instead of over-relying on scientific norms or uncritically adopting standard patterns of communication, narrative webs invite us to consider each narrative feature as a rhetorical decision to emphasize, minimize, select, deflect, include, or exclude.

Scientific storytellers are an important component of the narrative web. They are the source of the story's content and infuse stories with credibility and trustworthiness. Conversely, a story's content can shed a negative light on the storyteller and make them a target of confusion and rage. Rival storytellers play a prominent role in undermining science storytellers and proposing that scientific stories are incomplete and incompatible with their own stories. Many science storytellers seem aware of these shortcomings and acknowledge that there are competing stories that undermine their own.

Given that stories are always already embedded in other stories, how might we alter science's storytellers and content? One valuable lesson we learn from rival stories is what content is relevant to people, so that we may tailor science's stories to address these concerns or make links between scientific information and those concerns. Additionally, rival stories remind us of the nexus of concerns that people weigh before taking action.

Disingenuous rival stories clearly communicate that we cannot rely on all individuals having an instinctive trust in science or believing that "authority" and "expertise" is made up of just a title. By considering the role of storyteller and content, some science storytellers may reflect on their own role as communicators, seek out co-storytellers or alternative storytellers, or adjust their content to better connect with audiences and reduce concerns about the trustworthiness and relevance of the information. Instead of relying on a traditional notion of credibility and narrowing the topics covered, science's stories can strategically borrow from and collaborate with rival stories.

5 Constellation Practices

Don't Look Up, directed by Adam McKay, premiered on Netflix in late 2021. The film stars Jennifer Lawrence and Leonardo DiCaprio as astronomers Kate Dibiasky and Randall Mindy, who discover a comet on a crash course for Earth. After reaching out to US President Orlean, played by Meryl Streep, Dibiasky and Mindy are stunned to learn that she is more concerned about how the impending meteor will play to audiences during the election cycle than its catastrophic impacts to the planet. The global population becomes divided on the urgency and severity of the comet. Once it is visible in the sky, the "Just Look Up" campaign encourages people to observe with their own eyes the impending danger. The "Don't Look Up" campaign, alternatively, calls on people to ignore the comet as an anti-economic scare tactic. A technology billionaire, Peter Isherwell, opposes destroying the comet because of the potential economic benefit of the rare Earth minerals it contains. Isherwell's plan to mine the comet fails and it is then too late to derail the comet's collision course. While many elites, including Orlean, escape on a space shuttle, Mindy, Dibiasky, and their friends and families share dinner together as the comet makes impact. Although the film ends tragically with the destruction of Earth, there is a nod to the hope that our relationships provide in the ending scene's communal gathering around food.

Don't Look Up set a Netflix record for most viewing hours in a release week and totaled 360 million viewing hours during its first month on Netflix.[1] The film garnered mixed reactions, earning a 56 percent from critics and 78 percent from audiences on Rotten Tomatoes.[2] Wikipedia classified the film as an "apocalyptic political satire black comedy."[3] The satirical aspects could in part explain the mixed reactions; some felt the satire was too "on-the-nose," while others thought it hit the perfect mark regarding the current state of science communication and its obstacles.[4] *Don't Look Up* is about a comet, but the metaphor is a thin reference to climate change. Media and communication scholar Julie Doyle argues,

> By using satire to reveal the inadequacies of present socio-economic and political systems to acknowledge scientific evidence of the comet and respond to its extinction level threat, the viewer [of *Don't Look Up*] is called upon to make the associative link between the fictional comet and that of the global climate crisis.[5]

Doyle praises the film's critiques of extractive capitalism and quick "techno-fixes," but points out that in satirizing the current situation, the film centers White masculine voices and decenters "equity and justice" as global solutions to the climate crisis.[6]

We need more stories that provide new visions rather than recreate and satirize old ones. In *Beyond Climate Breakdown,* environmental communication scholar Peter Friederici argues that we should replace our capitalist stories of greed with comedic stories of hope.[7] Science communication is certainly not the same as entertainment media, but similar lessons about varying the stories we tell and not relying on the same tropes and patterns of the past can help guide us to create more engaging and interesting stories for audiences to learn and care about scientific content.

One way to vary our stories is through constellation practices, which refer to the strategies and considerations science communicators have available to them to build diverse shapes on the narrative web structure. Instead of prescribing a particular shape or even ideal wedges to be mapped on the hub spiral, constellation practices invite flexibility and adaptability. I do recommend at least two narrative features be mapped on the "hub spiral" of the web to anchor the story in specific people, places, and time periods and with credible storytellers and relevant content. Other features can be mapped at the meso-ring and macro-ring to create constellation patterns as best suit a storyteller's goals, circumstances, and audiences.

In this chapter, I outline six strategies that science communicators can use when they want to use any of the narrative features as hub wedges. These strategies are (1) humanizing science by using scientists as characters, (2) empowering audiences by positioning them as active heroes, (3) shrinking expansive scopes by providing specific examples, (4) managing uncertainty by using concrete language and visuals, (5) locating strategic overlaps between science and rival stories, and (6) decentering science. I additionally offer two strategies for experimenting with meso-ring and macro-ring mapping across the wedges: (1) using characters as bridges to larger entities and actions, and (2) working across sequence and scope rings through inductive reasoning and metaphors.

CHARACTER AND ACTION AS HUB WEDGES

To use character and action as hub wedges, I recommend that science communicators humanize science by including scientists as characters and empower audiences by positioning them as active heroes. In experimenting with constellations, science communicators can also consider using characters as bridges to larger entities and actions to incorporate meso-ring and macro-ring characters and actions.

Micro-ring character and action wedges can help create engaging stories. Political scientists Michael Jones and Holly Peterson argue that telling stories with clear heroes is effective for climate communication because it creates positive associations with the heroes and fosters negative perceptions of the villain.[8] Previous research I conducted with Angeline Sangalang found that stories about climate change that were "morally clear" were more engaging across the political spectrum than stories that were morally ambiguous.[9] Creating hero and villain characters, and fostering connections between audiences and the hero, can draw battle lines in ways that promote action and engagement, especially on politically roadblocked topics such as climate change.[10] Centering people, both scientists and public actors, as characters can increase audience identification with characters and help the audience connect more deeply to the story.

Humanizing Science

Scientists tend to evoke an objective, unbiased, and distanced relationship to their science, where they and other human influences are removed in service of "neutrality." A norm of scientific discourse is to use the passive voice and remove scientists themselves from the process of science due to "professional norms that encourage them to focus on the object of their research."[11] To say that science occurs, but with no person orchestrating or designing the science, is to remove the human part of the scientific process, which may be less appealing to audiences and may foster a lack of responsibility on the part of scientists for their participation in science.[12] This tendency to downplay human influence can harm the creation of rich characters that audiences can follow, connect with, and remember.

Instead, science communicators and scientists can locate themselves and others within the story. In doing so, we endow scientists and other science communicators with the qualities of scientist citizens who act on behalf of science *and* the public good.[13] Including scientists and science communicators as characters can accentuate the human element that underlies all scientific endeavors and help establish trust for scientists as characters and as storytellers.

Communication scholars Jean Goodwin and Michael Dahlstrom argued that a scientist's credibility is made up of a variety of features, including expertise but also personality and vulnerability. They propose that climate scientists in particular should engage in conversation with "doubtful and dismissive audiences, undertak[e] burdens of proof," and empower "audiences to assess the science themselves" in order to build trust with audiences.[14] Scientists who perform these actions express vulnerability, which Goodwin and Dahlstrom argue is important to establishing trust because "a person who trusts is vulnerable twice over" by opening themselves up to being wrong and to being "betrayed" by the person they put their trust in.[15] In other words, science storytellers must be vulnerable in order to expect that vulnerability in return.

One way that science stories can be vulnerable is to humanize their storytellers, make them more prominent, and open them up to questioning and conversation. The introduction to the storyteller and content chapter illustrates some of the potential risks of such advice, especially for those who are members of marginalized groups. By opening oneself to those potential risks, science's storytellers can build trust and connection between audiences and the stories' sources, thereby generating vulnerability. Oceanographers Elisha Wood-Charlson and colleagues propose recentering the scientist to "humanize the research by *telling the story of their experiences*" and suggest that "authors weave themselves into the action of story, even if just slightly."[16] Stories of scientific discovery about and by scientists can be incorporated into science education at all levels in order to encourage early and continued engagement with scientists as present and active within science's stories.[17]

Furthermore, recognizing the role of researcher within knowledge production embraces an Indigenous perspective on methodology. Shawn Wilson argued that researchers

> have to be true to yourself, and put your own true voice in there, and those stories that speak to you. That is retaining your integrity; it's honoring the lessons you've learned through saying that they have become a part of you.[18]

Part of engaging in Indigenous methodologies is to reject the veil of objectivity defined by Western science and instead to see the researcher as an important character within the research story who shapes and interprets it. Wilson further argued that information from personal stories of people's life experiences is "internalized in a way that is difficult for abstract discussions to achieve."[19] By moving the storyteller into the narrative and humanizing science's characters, we can voice science communication on the micro-ring.

Humanizing science's stories can be difficult depending on the topic. The stories of evolution, understandably, have few human agents, and we should not simply insert human characters into our past for the sake of adding them. Any additions or modifications to science communication based on narrative webs should still be scientifically accurate. One strategy for humanizing evolution's stories is to use scientists researching the topic as a frame through which to discuss evolutionary science. Consider Charles Darwin's canonical text *On the Origin of Species,* which was written as a diary. Describing the text, rhetorician of science John Angus Campbell argued,

> Darwin's opening chapters present natural selection and the ideas leading up to it not as abstract concepts or even as hypothetical possibilities, but as concrete experiences which are to strike the reader with the same force and directness with which Darwin himself was struck by his initial vision.[20]

On the Origin of Species brings the audience along on the journey to discovery, fostering identification between the reader and Darwin. Science's stories can thus highlight the scientists behind the experiments and discovery, treating them as people who are using science as a process to achieve new knowledge.

In the book *Undeniable,* Bill Nye described the journey of discovering the fossil Tiktaalik, a fish that has fins much like legs, marking the transition between sea-dwelling and land-dwelling life. Nye wrote,

> Scientists led by the tireless Neil Shubin of the University of Chicago found exactly such a fossil swamp in northeastern Canada, and went there and found Tiktaalik. Think about it, an ancient animal was surmised to have once existed. Researchers figure out where it would have lived. They went there and proved it. Amazing.[21]

Instead of stating "Tiktaalik was discovered," Nye writes that Shubin and a team of scientists went on a journey and discovered it. While Shubin is centered, this narrative also recognizes the research team that accompanied Shubin on Tiktaalik's discovery. This brief paragraph evokes heightened drama of discovery and the adventure of researchers on a scientific pilgrimage that confirmed a great mystery of evolutionary science.

Climate Outreach publishes case studies from Intergovernmental Panel on Climate Change scientists demonstrating how they connect to communities and translate global data for local use. Dr. Aïda Diongue Niang's story is about conversations in Senegal, where locals were aware of dramatic climate changes, but they did "not necessarily know why" they were happening.[22] Forgoing formal slideshow presentations, Niang set up informal open lectures to explain why changes were happening and how future changes could impact the community. In her story about speaking about climate change in Indonesia, Dr. Intan Suci Nurhati noted that it was important "to take my scientist jacket off and understand my audiences' needs [for] seeking my expertise."[23] Instead of creating separations between the technical and public spheres, Nurhati humanizes herself and connects to the nontechnical needs of the communities she is speaking with. These narratives demonstrate how scientists can be characters in their own stories and humanize themselves for audiences, thereby building "trust" through conversation and "codesign" of solutions.[24]

There are a few caveats when humanizing scientists in science communication. One is that the focus on individual scientists should not be at the expense of recognizing collaborators and teams who are often involved in scientific study but may not get the recognition of the lead, and oftentimes White male, scientist.[25] Care should be taken to acknowledge the many people who may be involved in scientific discovery to avoid the image of the lone scientist that may distort how science is actually conducted. Even in acknowledging the various positions involved in research, centering a few figures within a story about a research team can help humanize the scientific process and provide a few minds for the reader to connect with and follow.

When making choices of who to include as characters, science's storytellers should consider diversifying the scientists represented to show that scientists are women, people of color, queer, disabled, Indigenous, intersectional, and members of other marginalized groups. In so doing, naming various people and communities who are actively involved in the process of science can encourage the recognition of a variety of diverse voices and bodies in the discipline. In *The Disordered Cosmos,* astrophysicist Chanda Prescod-Weinstein explains concepts in physics and astronomy while reflecting on her personal journey through academia, her struggles, and the many obstacles to diversity and representation in the field. By emphasizing the people behind a variety of scientific practices, we can not only give scientific stories specific characters and actions to connect with, but we can also increase awareness of science's diversity and challenge White, cisgender, heteronormative, patriarchal constructions of "science."

Empowering Audiences

Engaging members of the public as valuable members of the scientific process is a powerful constellation practice. Defining characters only as formal scientists or scientific institutions ignores the voices of individuals with firsthand knowledge of environmental changes who may not be recognized within the formal bounds of "science." One way to include more community members in science's stories is to encourage citizen science engagement. Citizen science projects make the public cowriters of and active characters in science's narratives through collaborative scientific research. Much like scientist citizens are members of the technical sphere who also do work in the public sphere, citizen scientists are primarily members of the public sphere but who actively participate in the process of science. If people resonate with characters or *are* characters in the narrative, they can envision themselves as having positive effects on the story's outcome.

One of the earliest citizen science projects was "SETI@home," where people offered their computers' background capacity to process data to help the Search for Extraterrestrial Intelligence.[26] The Citizen Science Association website hosts two thousand citizen science projects that are engaging one million volunteers.[27] For example, the Cornell Ornithology Lab invites members of the public to upload information about bird migration behaviors in their backyards, so the patterns can be aggregated to reach conclusions about how climate change is affecting bird migration nationally.[28]

Empowering audiences does more than influence the character and action wedges. Because an individual is highlighted, a story's scope can also become more concrete, actions more specific, and content more relevant. Compare, for example, the narrative web of an individual's participation in a Citizen Science Association project from the Cornell Ornithology Lab with the larger narrative framework of climate change's stories. Based on analyses from the previous three chapters, we can consider climate change's stories from institutionalized sources, in aggregate, to have large scopes, future-oriented sequences, and broad or unnamed characters and actions, with content somewhat removed from audiences' frames of reference, and mixed trust in its storytellers. Alternatively, a

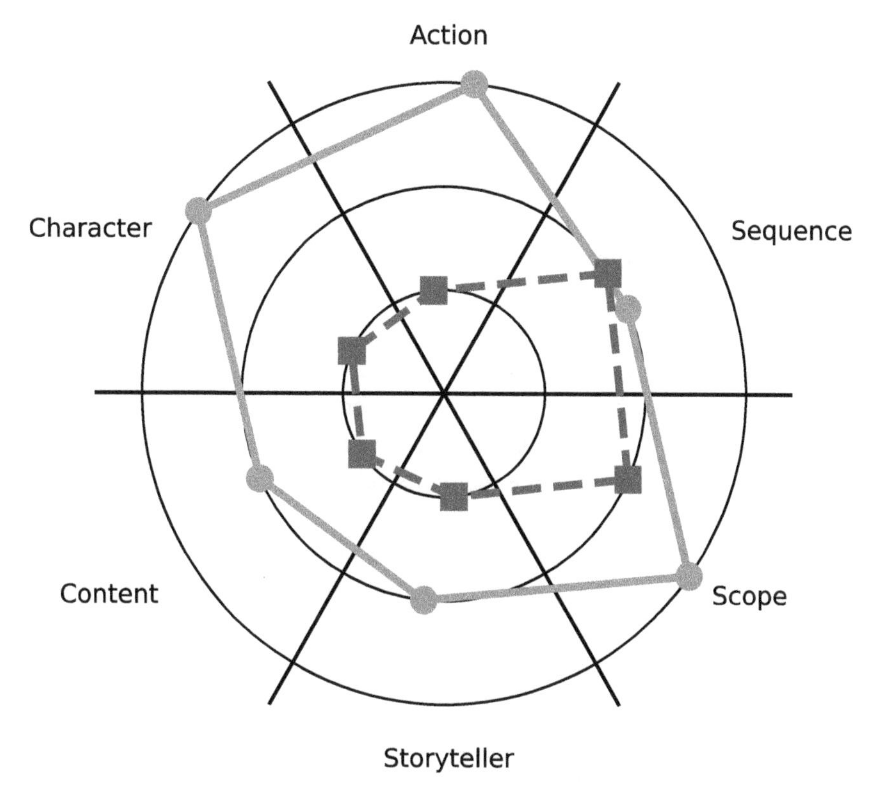

Figure 7. A narrative web showing the collective, institutionalized narratives of climate change mapped using a solid light gray line with circles, and an individual's Citizen Science Association experience mapped using a dashed dark gray line with squares.

narrative about an individual's experience in a citizen science project narrows the scope to community and regional bird migration patterns, connects to the individual's interests, and places them within the narrative as a character performing concrete actions. In the narrative web, the two constellations can be seen as separate stories while also simultaneously as parts of circulating climate discourse (see figure 7). Through citizen science, climate change is not a far off, abstract concept but a tangible, present issue that people can measure and see in their community.

Citizen science can also happen in classrooms. In its *Thinking Evolutionarily* report, the National Research Council encouraged classrooms to position students as scientists. The report recounted a science teacher's experience in their classroom where students mimicked Darwin's experiments on evolution: "Students are testing evolutionary theory with data, and they have the pride of ownership of their investigation and their products. . . . They are investigators. They're coming to learn science and do science."[29] Science classrooms should be spaces for experimentation and discovery and enable students to see themselves following the journey from science's beginnings to its endings in new knowledge.

It may not always be feasible to engage audiences in citizen science, so mediated versions can be another option. Through interactive media like gaming, audiences can be active players in science's narratives. Games' interactivity means that the story is not closed or predetermined, making players a type of coauthor to the narrative. Communication scholar Aaron Hess specifically discusses the elements of immersion and first-person perspective as unique to gaming, in which players have a sense of control and "direct contact" with the narrative's contents.[30]

From basic choice-based games such as Plasticity and the Climate Game to the more complex worlds of Highwater, climate games are becoming more popular.[31] Daybreak is a 2023 cooperative game by Matteo Menapace and Matt Leacock, the creator of Pandemic, in which players take on the role of world powers who are "building a tableau of technologies and policies to protect communities and decarbonize their economics."[32] While the game is still forthcoming at the time of this writing, one beta tester wrote that "the escalation of disaster with increased temperature means that every tree etc. is so important and something planted in round 1 will continue helping thereafter, and this reminds us to act sooner rather than later."[33] Interactive games can be one way to help audiences see themselves as characters in science's stories and feel empowered to test out solutions and gain valuable insights regarding the climate crisis.

In addition to citizen science projects, classroom spaces, and interactive media, we can empower people's relationship with science by sharing the stories of individuals with direct experiences. Our Climate Voices collects "personal climate stories" to "humanize the climate disaster."[34] By highlighting stories from people already being affected by climate change, drought, and extreme weather patterns, we can put a face to those already being hurt by climate disruptions.[35] Similarly, Voices for Vaccines hosts personalized stories from parents and guardians under the heading "Why I Choose." These narratives are often accompanied by photographs of people and their children and rich descriptions of why they choose to immunize their family. For example, Genevieve from Albany, New York, notes that she chose "TDaP to reduce my baby's risk of pertussis. There are enough things right now that I cannot prevent. I'm grateful that pertussis isn't one of them."[36] In these stories, the science of vaccination and its health benefits is communicated through a parent, who is a character empowered to act instead of feeling compelled to.

The Centers for Disease Control regularly includes first-person narratives in its vaccination information materials, often evoking the voices of parents and guardians. For example, a whooping cough information sheet shares a statement from someone who is pregnant: "The whooping cough vaccine I got during my 3rd trimester will help protect my baby starting at her first breath."[37] Additionally, this information sheet addresses the reader using "you" language, encouraging them to be in a position of power as a decision-maker, fostering identification between the audience and the narrator. The sheet notes,

> During the third trimester of your pregnancy, you'll pass antibodies to your baby that will help protect her from this disease from the time she's born. These antibodies will last for the first few months of her life, when she is most vulnerable to serious disease and complications.[38]

Positioning parents and guardians as powerful protectors over their children's health inserts them back into the narrative as heroes. Instead of being replaced by governmental and medical authorities, parents can be powerful coagents in protecting their family's health.

We map scientific stories on the micro-ring of the character and action wedges by locating human characters within their stories—whether scientists, citizen scientists, or others—and empowering audiences to take positive, hopeful action in response. Instead of removing characters altogether, using abstract or collective characters, positioning audiences as blameworthy, or affecting their agency, science's stories can embrace a variety of specific, empowered characters. These characters perform concrete actions, function as someone to care about, and create a personalized journey with which people can connect and identify. Additionally, inviting audiences themselves to be part of stories or having them connect to stories can infuse a sense of hope, empowerment, and agency. This is not to say, however, that all scientific stories need to have micro-ring character and action wedges to be effective or to connect with audiences; it is one potential tool among others. Experimenting with other rings on the character and action wedges can open creative possibilities for science storytellers.

Meso-Ring and Macro-Ring Characters and Actions

Science's stories can incorporate meso-ring and macro-ring characters and actions to create more varied constellations within their narrative webs and to engage audiences on multiple levels. In terms of meso-ring and macro-ring characters, institutions and large international bodies can confer a sense of credibility and reliability in ways that individuals cannot. Scientific groups have authority in part due to their collective status that represents an agreement or common pursuit upheld by the implied many individuals who are part of those institutions.

There is evidence that teaching people about the scientific consensus on climate change may be a gatekeeping belief to increase acceptance of human-caused climate change.[39] However, a statistic such as a 97 percent consensus may be hard to conceptualize. In April of 2014, *Last Week Tonight with John Oliver* featured a segment called "A Statistically Representative Climate Change Debate." Instead of a media broadcast with two talking heads representing each "side" of the climate debate, John Oliver brought on three people to represent climate hesitation and then filled the set with ninety-seven other people in lab coats to represent the scientific consensus. Partially a satirical stunt, the segment illustrated in more concrete terms what scientific consensus looks like. Political communication scholars Paul Brewer and Jessica McKnight measured the effects of viewing Oliver's segment and found that it increased participants' climate change beliefs and their beliefs in the scientific consensus.[40]

Given the potential persuasive power of consensus, science's storytellers may incorporate meso-rings by appealing to collective characters. Similar to Our Climate Voices and Voices for Vaccines that center individual voices, the National Center for Science Education publishes

"Voices" documents that aggregate organizational statements. *Voices for Climate Change Education*, for example, is a document that gathers support for teaching climate change in education. The document notes, "Many leading scientific and educational groups have expressed support for teaching climate change in college, high school, grade school and informal settings."[41] Currently, the document contains sixteen statements, including from groups such as the National Science Teachers Association, the American Meteorological Society, and the United Nations Educational, Scientific and Cultural Organization. The National Center for Science Education also publishes *Voices for Evolution,* which includes hundreds of statements from groups supporting evolutionary science, such as scientific and scholarly organizations, educational organizations, religious organizations, and civil liberties organizations, which reflect "the full diversity of organizations and perspectives in support of teaching evolution in the public schools."[42]

In some instances, it may be cumbersome to name scientists or individuals involved in scientific research. Broad statements aiming to discuss what "the science says" or the current state of research may find listing scientists or providing backstories of affected individuals impractical. Additionally, individual characters and actions may be distracting to the larger scientific point. If science storytellers primarily use meso-ring and macro-ring characters and actions, they might instead consider employing scope and sequence as hub wedges. While I recommend at least two wedges be mapped on the hub spiral, having all of them mapped there would not resemble a varied constellation—readers may recall CJ's solely micro-ring narrative web from the introduction chapter. Therefore, storytellers may make strategic decisions regarding which wedges to map on the hub spiral or whether to combine rings on the same wedge, such as by including specific characters and consensus statements, or individual and community level action, in the same story.

SEQUENCE AND SCOPE AS HUB WEDGES

In attending to sequence and scope, we can help make science's stories more accessible, understandable, and relevant to people's everyday lives.

Micro-ring sequences create clear causal relationships between events chronologically through time. Micro-ring scopes operate within the spatiality and temporality of people's direct experiences. When using sequence and scope as hub wedges, science communicators can shrink expansive scopes by providing specific examples and placing the scientific topic in local and present contexts, and manage uncertainty by using concrete language and visuals.

It is important to remember that full deflection of broad temporal and spatial scopes would truncate science's discoveries and could lead to scientific inaccuracies. Additionally, concrete sequences may be overly limiting, restricting space for experimentation with the arrangement of narratives and inaccurately presenting the certainty of science's findings. Thus, science's storytellers may consider incorporating more abstract sequences and scopes depending on the topic, context, and audience. To experiment with meso-ring and macro-ring sequences and scopes, I recommend working across rings through inductive reasoning and metaphors.

Shrinking Scopes

One way to modify science's stories to account for expansive scopes is to narrow the circumference of the story to a more manageable frame of reference in terms of time (i.e., present and recent past) and space (i.e., local). In reducing the circumference, or what is contained within the scope of the story, we may define a story's temporal and spatial dimensions more narrowly based on places and time periods that are "second nature" to us.[43] Similar to the idea of using characters that are similar to audiences, shrinking scopes and creating localized examples "puts the audience in a frame of mind where they may be personally implicated in the topic" and thereby invites personal connection with the story.[44]

Narrowed scopes shrink the "distance" between audiences and a topic, such as climate change.[45] This is incredibly important considering that studies show people see climate change risks as "less serious," "less dangerous," and mostly affecting others more than themselves. Narrowing the spatial circumference of a story can be as simple as reducing national and international topics to a local focus. For example, communication scholars P. Sol Hart and Erik Nisbet noted that shrinking the "social distance," or perceived similarity and proximity, between the audience and the characters of climate stories reduced political bias and encouraged environmental activism.[46] By locating stories in communities and with people that audiences care about, we can encourage beliefs and behaviors in alignment with science's stories.

When telling the stories of climate change, science narrators can evoke local communities instead of expansive, global ones. Environmental humanities scholar Elizabeth DeLoughrey argued that there is an inherent irony in Anthropocene discourses that claim to operate on "the scale of the planet," but actually focus on limited perspectives that exclude marginalized voices and local communities.[47] In other words, attempts to universalize the telling of climate change's stories merely work to erase differences and laud certain perspectives (i.e., Western, patriarchal ones) as the de facto story. Telling specific, "provincialized" stories, DeLoughrey argues, can be a form of counterstory to the limiting practices of scientific discourse and thereby disrupt practices of demarcation and exclusion.[48]

In the comic book *Mayah's Lot,* the Center for Urban Environmental Reform tells the fictional story of a student of color, Mayah, who is facing an instance of environmental injustice in her own backyard.[49] An empty lot that Mayah wants to grow a garden in has been selected for toxic dumping. At the beginning of the comic, Mayah narrates the importance of environmental justice in a local context, saying, "Whatever [environmental justice] is, most people don't know until it's starin' them in the eyes. I had no idea that it actually affects every one of us, that is, until it came to my home." Instead of an abstract definition of environmental justice, *Mayah's Lot* uses a place-based narrative to communicate how communities of color are impacted by development decisions. The story is told in a local and present scope, but also supports the idea that environmental justice is a community-level activity, as Mayah recruits the help of many others to help her defeat the dumping plan.[50]

Scientific organizations can take cues from local stories by focusing on impacted communities and vulnerable populations. For example, the National Centers for Environmental Information publishes "State Climate Summaries," which tailor climate information to each state. Emphasizing unique local features and impacts can locate broad spatial scopes, such as the global scene of climate change, in audience's backyards. Getting even

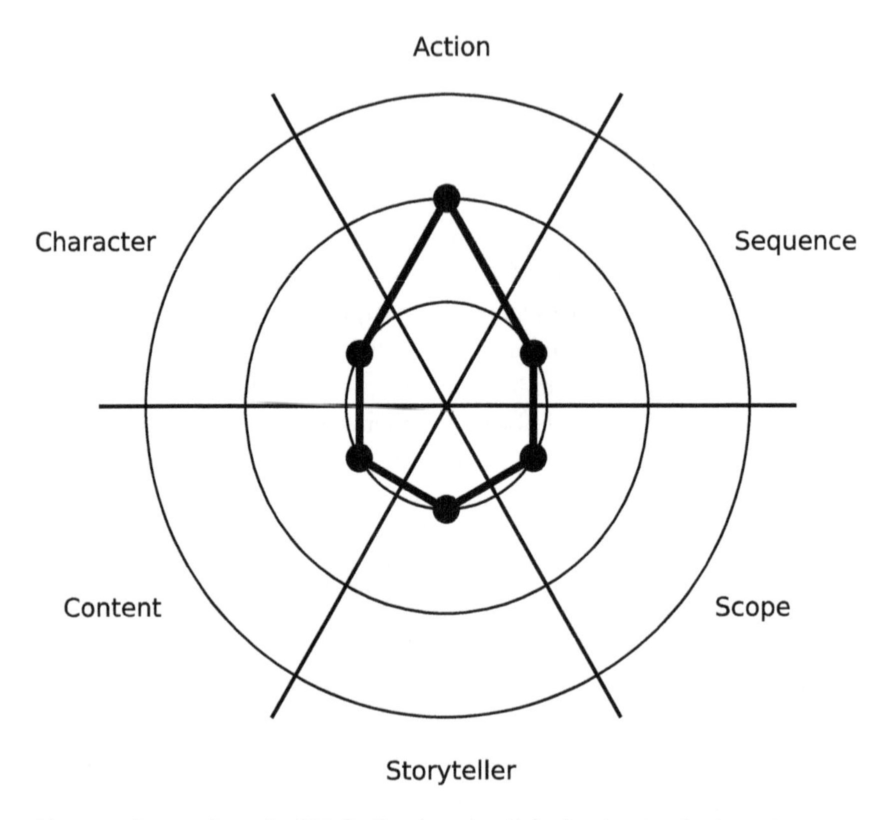

Figure 8. A narrative web of Nadia Ramlagan's article about water shortages in Nevada as told by members of the Pyramid Lake Paiute Tribe and the Fallon Paiute-Shoshone Tribe in Nevada, showing specificity throughout the narrative web and the action wedge mapped on the meso-ring.

more specific than statewide scopes, scientific organizations can focus on local communities and their needs. For example, the American Association for the Advancement of Science publishes news stories related to local climate change effects. Contributor Nadia Ramlagan's article and accompanying video discussed the lack of water available to the Pyramid Lake Paiute Tribe and the Fallon Paiute-Shoshone Tribe in Nevada.[51] The video and article pair first-person experiences from tribe members and representatives with scientific information about snowpack and the water cycle to paint a picture of climate change's effects on Indigenous tribes in Nevada (see figure 8).

The narrative web of Ramlagan's article has a local scope with specific characters narrating their own stories, discussing their firsthand knowledge of water scarcity that will affect not only Native communities but others in Nevada as well. Although the story does not propose specific actions that should be taken regarding the water shortages, it does raise awareness about the problem and implicitly call for some action to be taken.

Vaccination and COVID-19 storytellers can describe disease spread and vaccination rates at local and community levels. As a way of communicating scientific information, the *New York Times'* COVID-19 dashboard allows people to expand or shrink the scope of reported data by location, time frame, and type of data. From the entire globe to the United States to a specific county, people can see how close the virus is and calculate potential risks. In terms of temporal distance, users can select to see cases for the past two weeks up to "all time," thereby adjusting temporal proximity. Information about the present or the recent past creates more engaging narratives for audiences across the political spectrum.[52] Having stories in the present or recent past also avoids the natural uncertainty embedded in future-oriented stories, which will be discussed in the next section.

Across all case studies, we can adopt the strategy of shrinking scopes and narrowing circumferences by tailoring scientific narratives to local communities, for example, by providing vaccination statistics at the local and state level, focusing on climate change's effects on specific communities, and providing examples of evolutionary change in animal and plant life local to the audience. Shrinking scopes as a suggestion for science's storytellers emphasizes the importance of place-based scientific communication to engage audiences in concrete, meaningful experiences.[53]

Managing Uncertainty

Acknowledging uncertainty can both help and hinder science communication. Uncertainty is a natural part of the scientific process, so communicating it is both accurate and a way to build trust with audiences about what is known or unknown. However, communicating uncertainty through hedging or emphasizing what is still unknown opens the door for skeptics to demand 100 percent certainty and for the science to be "settled" before action can be taken.[54] One strategy for managing uncertainty is to adjust the language we use to communicate probabilities. In a 2021 study, authors Elisabeth Lloyd and colleagues, an interdisciplinary team of historians of science and scientists, proposed that scientists should not use the standards of the technical sphere to translate percentages into likelihood statements. Because scientific information is used for public decision-making, they note that scientists should use language aligned with standards of proof in legal and public policy contexts. For example, the authors noted that "climate scientists generally look for a probability of 90–100% before they call a scientific claim . . . 'very likely.'"[55] Changing the language of how we communicate uncertainty can frame climate change's consequences as more salient to audiences and align with more general understandings of probability.

Climate Outreach's *Uncertainty Handbook* offers another language shift by proposing flipping the language of uncertainty to risk. Instead of focusing on the uncertainties of climate models or predictions, science communicators can emphasize the likelihood of future risks through "clear practical examples of the risk of a village flooding, or a farmer's crops being destroyed, or a coastal building slipping into the ocean."[56] A narrative about a specific community's future risks, such as Niang's activism in Senegal or the effect of drought on Nevadan Tribes, can activate personal concerns and reduce social and hypothetical distance and thus uncertainty about the risks of climate change. In this sense, sequences that incorporate future uncertainty can seem more manageable through narrowed scopes and specific examples.

Another strategy to manage uncertainty is to tell stories that envision the future either with language or visuals. The famous prologue to Rachel Carson's *Silent Spring*, for example, visualizes a likely future as happening in the present if pesticides were not addressed. Multimedia and audio-visual mediums can also be effective tools for science's storytellers. In a series of studies, climate researchers Saffron O'Neill and Sophie Nicholson-Cole found that images can promote fear about the uncertainties of the future but also can inspire feelings of agency to change it. Feelings of agency emerged when participants saw images that "take account of individuals' personal points of reference" and included "locally relevant climate change imagery."[57]

In the case of vaccination and COVID-19, it may be harder to visualize diseases that operate on microscopic, aerosolized, and nearly invisible levels. Especially given the temporal distance many people feel from diseases such as measles and polio, the perceived side effects of vaccination may seem more tangible and worrying. Additionally, the sometimes unseen and unfelt presence and symptoms of COVID-19 may give some the impression that there is no real danger. Medical illustrators Alissa Eckert and Dan Higgins created the iconic Centers for Disease Control image that represented the SARS-CoV-2 virus that causes COVID-19 in bodies. Eckert recounts being asked by the Centers for Disease Control "to create a visual identity for the virus."[58] In discussing their choices of color, texture, and shape, the illustrators tell a story that makes viewers "feel like you can touch it [to bring] it closer to reality. It is a way to take something complex and abstract and make it tangible. We gave the virus a face."[59] The image brought "the unseeable into view" and invited people to reflect: "if they reach out to touch something, they'll picture that spiky gray blob and pause," lest it be introduced into their bodies.[60] The point of the image, therefore, is simultaneously to communicate scientific information about COVID-19 spreading across the globe and to translate a way of sensing and knowing the coronavirus visually.

Health communication scholars Andy King and Allison Lazard argued that while there are recommendations from health organizations such as the National Cancer Institute about using "culturally appropriate" and "attractive" visuals, these suggestions are relatively "vague" in terms of how one might select images that meet these criteria.[61] In using visuals, therefore, science's storytellers should tailor images, just as words, to the target audience and consider how images may be used to reduce uncertainty by shortening temporal, spatial, social, and hypothetical distances. For example, science storytellers may use images of communities and people similar to the target audience being affected by a disease or illness to increase perceptions of personal and community risk. Uncertainty can thus be reduced or managed through visualizations and concrete language that make the future seem more tangible and present.

Working across Sequence and Scope Rings: Inductive Reasoning and Metaphors

Working across the rings incorporates the micro-ring, meso-ring, and macro-ring on the sequence and scope wedges in concert. Instead of only using small scopes and micro-ring sequences, this strategy proposes starting from case studies to move audiences into larger, more expansive frameworks. DeLoughrey calls moving between abstract and concrete "telescoping," a method which emphasizes "how they mutually inform each other."[62] For example, one can engage in a "telescoping of the past through the present" by using history "as a means of figuring a contemporaneous moment of crisis."[63] We can enact such telescoping through narrative webs by building stories that put the abstract and the specific in conversation with one another, thereby mapping across the three rings within the same story.

Some scholars caution against overemphasizing a local focus. On the topic of climate change, for audiences who already are experiencing climate change or conceptualize it as a local, immediate concern, micro-ring scope messages may do little to raise awareness or motivate action.[64] If one were hoping to increase information about climate change's relevance to an audience who may be skeptical of climate change's impacts, micro-ring scopes may be effective. For others, messages that raise the profile of the issues may be more effective. In my first book, I spoke with a group of religious environmentalists I called "harmonizers," due to the compatibility they saw between their faith and environmental advocacy. In my conversations with this group, I found that messages about the global impacts of climate change encouraged feelings of "strong activism," which emphasize enacting societal-level transformations instead of relying on moderate, individual changes.[65] What may be more helpful than focusing on climate issues and impacts is modeling the activism of others in the stories we tell, thereby increasing audience agency, such as through fictional narratives like *Mayah's Lot,* discussed in the "Shrinking Scopes" section of this chapter, or true stories of activism.[66] These are examples that can then be scaled up to larger generalizations about the importance and prevalence of climate action.

The strategy of moving from micro-ring examples to broader meso-ring and macro-ring ones can be conceptualized through inductive reasoning, or reasoning from specific examples to generalizations. Scientific experimentation operates through inductive reasoning in that specific experiments in aggregate inform larger principles. Similarly, the presence of various examples creates evidence of larger trends that can then be applied to groups more generally. Ramlagan's article about Indigenous communities experiencing drought argued, "Nevada makes a good case study for predicting how communities that depend on snowpack for water supply—50% of the global population—will fare under a changing climate."[67] Addressing drought in Nevada is thus important both because of its already present effects on Indigenous communities and the ability of Nevada's experiences with drought to represent likely issues in other communities. This powerful interplay between micro-ring and macro-ring scopes demonstrates the potential value in working across rings.

O'Neill and Nicholson-Cole found that the images that made climate change most important to participants were images that showed "dramatic visions or human or animal suffering at *both local and global scales,*" such an image of a "house falling off a cliff" and "flooded houses and people in Bangladesh."[68] Using local and global scales together, or telling stories *glocally*, can emphasize both the direct impacts and broad reach of climate change. Along these lines, Dylan Harris recommends that climate storytellers "find ways to contextualize [their] research" by connecting to specific communities before bringing "the universal concerns of climate change—rising global temperatures, for example—into focus, [thereby] holding the contextual and universal in tension."[69] In amassing examples of how climate change has affected various cities around the world, Al Gore's public presentations work across the rings to move inductively from examples to the conclusion that there is a global climate crisis. DeLoughrey analyzes the artwork *Mar Caribe* of Dominican artist Tony Capellán, which is a "collage" of salvaged flip flops that have had their straps replaced with barbed wire. *Mar Caribe* telescopes from the individual who lives in "barrios pobre" (poor neighborhoods) in the Caribbean and is made invisible and disposable by global capitalism.[70] Telescoping between specific examples and broader implications can invite recognition of communal effects and shrink distances between audiences and the relevance of scientific information.

In addition to working across scope's rings spatially, science's storytellers can use contemporary examples to scale into deep time frames of the past. In *Thinking Evolutionarily*, the National Research Council describes a teacher's evolution lesson where students start "with familiar examples from the present and recent past and gradually [work their] way toward the distant past."[71] Quoting the instructor, the report notes that students "can see that the exact same concepts and things that they know and can understand in the present or in the recent past apply to the ancient past." This process works to combat the difficulties that students have understanding large scopes of time:

> Once [students] have used mechanisms to help them understand the depth of time we're talking about, and you start multiplying the kind of changes we see over short time to those longer times, they begin to understand how this can all work.[72]

In this strategy of working inductively, students start from individual examples that are relevant to them and then work backward into larger temporal scopes and more generalized understandings of Earth's history.

Specific examples can also be connected to the scientific topic at hand through the deployment of metaphors. Metaphors are artistic comparisons between a tenor and a vehicle. The tenor is the topic of discussion, and the vehicle is the lens through which the tenor is compared. In *Don't Look Up,* the tenor is climate change and the comet is the vehicle that helps audiences see the topic of discussion through a different lens. While there are many differences between a comet and climate change, their similarities as world-ending disasters create a useful parallel for engaging audiences on climate change's importance and urgency .

Metaphors in science communication can foster deep, rich understandings of scientific topics otherwise inaccessible through technical language.[73] It is important to note, however, that sometimes metaphors can complicate science communication, encourage the carrying over of inaccurate understandings, and may even lead to harmful attitudes and behaviors.[74] In using metaphors as tools, just as we use language and stories, we must be reflexive of the comparison being made and the associations that follow.

An example of the productive power of metaphors is astrophysicist Carl Sagan's cosmic calendar. Developed for the show *Cosmos,* the cosmic calendar "compresses the local history of the universe into a single year."[75] By comparing nearly fourteen billion years to a single year, people can better grasp the stretch of time between cosmic happenings and the relative time frame that humans have occupied. This is how the cosmic calendar was explained in an episode of *Cosmos*:

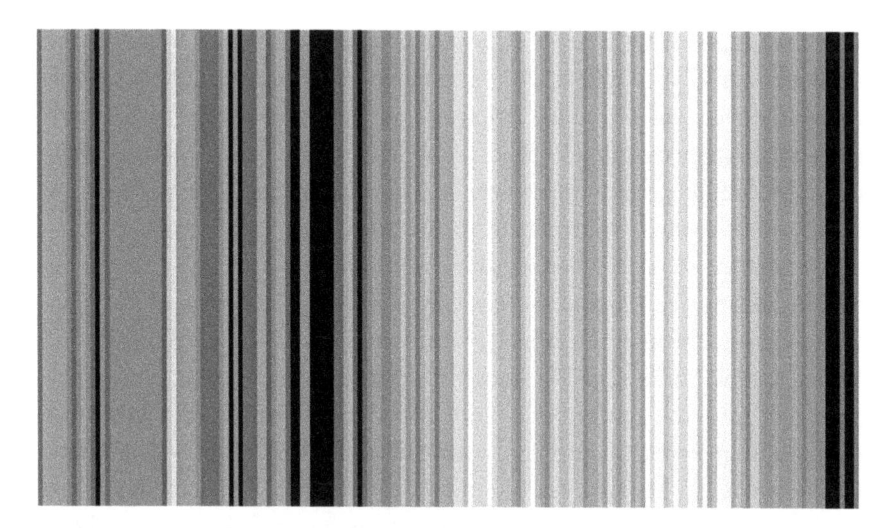

Figure 9. A visual display of temperature stripes representing annual average global temperatures from 1850 to 2021, with cooler years marked in shades of blue (grouped mostly on the left) and warmer years marked in shades of red (grouped mostly on the right). Created by Professor Ed Hawkins from the Show Your Stripes project, https://showyourstripes.info/.

> If the universe began on January 1st, it was not until May that the Milky Way formed. . . . Everything humans have ever done occurred in that bright speck at the lower right of the cosmic calendar. The Big Bang is at upper-left in the first second of January 1st. 15 billion years later is our present time: the last second of December 31st. Every month is one and a quarter billion years long. Each day represents 40 million years. Each second stands for some 500 years of our history; the blinking of an eye in the *drama* of cosmic time.[76]

Sagan illustrates the story of the universe's forming from the Big Bang to human life, creating an appreciation for the length of the universe's history through transforming it into a time scale—months, days, and seconds—that people can more readily understand than millions and billions of years.

Visuals can be deployed in the abstract by transforming graphically based data into aesthetically pleasing representations. The Show Your Stripes website represents years of warming and cooling through the colors blue (cooling) and red (warming) to show average temperatures in a location annually (see figure 9). Show Your Stripes creator Ed Hawkins writes that the graphs were created to be "as simple as possible" by removing numbers and "specific details."[77] Changing temperatures are communicated in a way where there is "minimal scientific knowledge required to understand their meaning."[78] Show Your Stripes plainly a micro-ring mapping that would provide more specific details and instead embraces the beauty and simplicity of graphical abstraction.

Metaphors about science can be communicated through the arts as well. In the musical *Hadestown,* the Greek myth of Orpheus and Eurydice is reimagined through the lens of climate change and industrialization.[79] Lured by the rumors of Hadestown being a land of plenty, Eurydice descends into the underworld and is trapped there by the capitalist machine that keeps people poor and tied to their back-breaking labor. Orpheus goes to Hadestown to rescue Eurydice and strikes a deal with Hades that Eurydice can be saved if he trusts her to follow him without looking back. Mirroring the ending of the original myth, Orpheus does look back and Eurydice is lost back to Hades(town). Despite this unfortunate loss, the musical ends with the opening number sung in reprise and the actors reenacting the first scene, indicating that the story has restarted and looped back to the beginning. The reprise lyrics read:

> Cause, here's the thing:
> To know how it ends
> And still begin to sing it again
> As if it might turn out this time[80]

Despite the ways *Hadestown*'s story has inevitably turned out in the past, the singing continues and the story restarts, the characters committed to trying once again to rewrite the story and have Orpheus save Eurydice and rebalance the underworld and overworld. Although more hopeful than the inevitable looping in *mother!* (discussed in the sequence and scope wedges chapter), the cyclicality of these stories inspires hope that change might happen and a new ending may be written if we try again to break the cycle.

The cyclical narratives of *Hadestown* and *mother!* embrace nonlinear timescales and make room for untidy endings and continuations in ways that closed narratives do not. The cosmic calendar or a comet crashing into Earth, like many metaphors, may give the impression that there is an ending to the flow of time, a tidy culmination that can be explained and expected. Education scholar Margaret Kovach (Plains Cree and Saulteaux) wrote that "linearity . . . is antithetical to the fluidity and cyclical nature of Indigenous epistemologies and methodologies."[81] Cyclical narratives are potentially more accurate to our understanding of certain scientific topics as endless and may be effective in capturing an audience's attention regarding the severity and urgency of issues such as the climate crisis.[82]

Metaphors invite audiences to make interpretive leaps to understand scientific topics such as climate change through different vehicles. The term *vehicle* itself is a metaphor for how features are carried over from one topic to another, implying movement and a channel through which movement occurs. We may also imagine audiences as vehicles by encouraging them to participate in metaphor through movement. The app *Deep Time Walk* is an interactive podcast that narrates a dialogue between the "scientist" and the "fool" from the beginnings of the universe to the present day as listeners walk through their communities. Asking walkers to put in their height, the app uses a phone's movement sensors to pace the listener through billions of years of time as the narration plays, with each meter walked equivalent to one million years of Earth's history. In total, listeners walk a little less than three miles (around 4.6 kilometers) with their movement tracking the evolution of life through the scientist and the fool's conversation.

Deep Time Walk narrows the scope of evolution's temporality in making the time frame of evolution physically felt and intimately connected to individuals, yet still retaining large numbers to illustrate how such vast changes were able to occur over time. The app notes that its goal is "to help foster wider understanding of deep time and to show humanity's deep ancestral history and interconnectedness with all life," thereby educating publics about "deep time," such as vast evolutionary timescales, by placing it in a more tangible framework.[83] Additionally, the app walks people through their communities and surroundings, asking them to pause and consider what they see, but also what they hear and feel as they imagine the universe and the Earth being created around them.

An understanding of deep time is also a principle of Indigenous worldviews. Instead of a Western science approach that tends to "study over a large landscape in a short time series," Indigenous sciences tend to emphasize "cumulative and concerted experience over a long time horizon in one location."[84] Fostering a sense of deep, emplaced time, therefore, can expand our understanding of what counts as scientific data to include ancestral and multigenerational knowledge. This is in line with the Iroquois concept of "Seventh Generation Stewardship," which emphasizes that current actions should take into account "the benefit of the seventh generation into the future."[85] Spanning nearly 200 years into the future, this perspective "ensure[s] that decisions made today would benefit the children of the future" by building sustainable worlds.[86] Pairing knowledge of the present, past, and future with tailored, personalized stories can help make the expansive scopes of science's stories more accessible and relatable.

STORYTELLER AND CONTENT AS HUB WEDGES

To use storyteller and content as hub wedges, science communicators should specify trustworthy storytellers and include relevant and important topics. I also recommend locating strategic overlaps between science and rival stories and decentering science. We can locate instances of unique opportunity, where scientific discourses engage productively and even collaboratively with traditionally hostile or otherwise competing narratives.[87] These moments may come in a variety of forms and can be taken advantage of and even produced by finding bridges between disparate spaces. Bridges are people, groups, and authorities who can carry over their credibility and bring audiences to science and scientific topics otherwise inaccessible or difficult to access.[88] Science communicators can meet people where they are at and work within topic areas that are already accepted and of interest to the audience.

If people resonate with rival stories, for example, due to a religious or political identity, it can be incredibly difficult to shift this perspective. Walter Fisher argued that "the only way to bridge this gap, if it can be bridged through discourse, is by telling stories that do not negate the self-conceptions people hold of themselves."[89] Consequently, the process of adopting new narratives requires at least partial integration with the existing identity or the construction of an overarching perspective that can include both the old identity and the new information so as not to compromise or negate people's already held beliefs.

Locating overlaps between narratives can help people be more willing to accept scientific stories without feeling that their identity is under threat. Instead of encouraging someone to occupy a completely different perspective, locating strategic overlaps encourages people "to see their dominant frame," such as religion, politics, and economics, "as compatible" with science's stories.[90] Strategic overlaps may be found in shared authority figures, common values and priorities, identity markers, geographic overlaps, and mutual interests. Once identified, these overlaps can spark new insights and reduce the perception of conflict between science's stories and rival ones. By championing people and places at the intersection of science and rival stories, we can foster science's credibility and its relevance to people's daily lives. There are many potential arenas for overlap, but I outline productive starting points in the rival stories of religion, politics, and economics.

I also explore the strategy of decentering science, which links scientific topics to other areas that may be more relevant to audiences or prioritizes other values, such as diversity and equity. While science is the primary component of communicating scientific information, this strategy refers to avoiding an overreliance on relaying facts, statistics, and the minutia of methods and scientific processes. Oftentimes, a lack of knowledge of scientific information does not drive, or does not reliably drive, science skepticism.[91] An alternative is to center the values, interests, and priorities of audiences. This strategy values and incorporates scientific information but positions it in a supportive and complementary role to shift other, more relevant content to the forefront of the story.

Religion

While some people view religion and science as compatible, others view them as separate and distinct ways of knowing.[92] To reduce perceptions that science and religion are incompatible, scientific stories can emphasize places of common ground. For example, the Understanding Science website noted that "the two institutions [of science and religion] deal with different realms of human experience. Science investigates the natural world, while religion deals with the spiritual and supernatural—hence, the two can be complementary."[93] This same website cited a survey that reported nearly half of scientists at top research universities reported having a religious affiliation and 75 percent "believe that religions convey important truths."[94]

In its *Voices for Evolution* report, the National Center for Science Education highlighted religious groups that agree with evolutionary science, such as the American Jewish Congress, the Episcopal Church, the Lutheran World Federation, the Roman Catholic Church, and dozens of other religious groups. In total, the third edition of *Voices for Evolution* has 176 statements from religious groups. These statements help break down "illusory correlations" that scientists are dogmatic atheists who are intent on destroying religion, as creationist groups such as Answers in Genesis would claim.[95]

In finding overlaps between religion and science, we can also locate storytellers who occupy multiple spaces. In the case of climate change, Dr. Katharine Hayhoe is an atmospheric climate scientist and Evangelical Christian who is outspoken on her dual identities and how they both motivate her to climate change advocacy.[96] Hayhoe's "global weirding" YouTube videos include scientific information about climate change and information about religion's role in environmental advocacy, thereby uniting scientific and religious content.

More broadly than to a single person, appeals can be made to Creation Care groups, who combine care for the environment with biblical mandates to serve God, protect the Garden of Eden, deny sin, and be pro-life.[97] In a survey I conducted of members of Creation Care groups, 100 percent of respondents found no conflict between science and religion, and instead reported that incompatibilities and inconsistencies between the two are largely a result of human misinterpretation.[98] Creation Care groups, such as Interfaith Power and Light, the Evangelical Environmental Network, the Climate Witness Project, and the Catholic Climate Covenant, consider faith and environmentalism to be intertwined commitments. Finding representatives from these groups and incorporating them into our science communication can encourage some audiences to rethink assumptions about a perceived antagonistic relationship between science and religion.

While religion is not a common source of rival stories to vaccination, it is important to note that some vaccination hesitancy is rooted in religious convictions.[99] Religious arguments against vaccination draw upon the sacredness of the body and the presence of substances in vaccines that may violate religious beliefs.[100] Some of these appeals are certainly genuine, but there is also evidence that religious appeals may be leveraged as a strategic resource to gain legal exemptions to vaccine mandates.[101] Because religion does play a role, albeit not as prominently as in evolution and climate change controversies, it is still a potentially fruitful strategy to incorporate religious arguments in vaccine stories. For example, calling upon the statements of religious leaders that encourage vaccination may associate religious authorities with support for vaccination and undercut claims that one's faith prohibits vaccination.[102] Religious leaders of Judaism, Islam, and Catholicism all note that vaccination does not violate the tenets of their faith, making religious leaders potentially influential distributors of vaccination messages.[103]

Across all four case studies, championing religious voices and highlighting agreement between religious thought and scientific conclusions can make science's stories stronger. Scientific storytellers are more credible and trustworthy if they are not viewed as anti-religious or as challenging people's religious beliefs. Wilson argued that academia and the process of Western research are typically "devoid" of any mention of spirituality, even though spirituality "is an integral, infused part of the whole."[104] Although scientists may want to strip the supernatural, spiritual, and nonempirical from experiments and interpretations, for members of the public, their absence may be concerning or confusing. Excluding fundamentalist interpretations that deny the basic facts of science, there is considerable overlap and common ground between religious ways of life and mainstream science.[105] Focusing on these overlaps instead of their differences can reduce incompatibility with rival stories, increase the credibility of science's storytellers, and create stories with high narrative fidelity for religious audiences.

Economics and Politics

A common theme in disingenuous rival stories to science is the presence of hierarchies that privilege the economy over other concerns, such as public health or the environment.[106] When money and markets outweigh other concerns, it can be difficult to make people care about other content areas, such as scientific topics. Consequently, working with economic content can be a strategy to bridge gaps and connect scientific topics to financial priorities.[107] Appeals to the overlap between economics and science is most relevant in the controversy over climate change, where neoliberal economic policies work against environmental protection and long-term environmental planning for the benefit of short-term economic gains.[108] Money is typically the standard by which to measure environmental actions *and* a reason to doubt that climate scientists are unbiased.[109]

There are, of course, concerns and risks in "embracing a neoliberal ideology to talk about the environment," but there are also potential opportunities because working within frames can open lines of dialogue without "dismissing [people's] frames out of hand."[110] Philosopher Kevin Elliot argued that "indirect" arguments can be made for environmental protection by emphasizing "economic growth, better healthcare, and more secure jobs."[111] For example, we can tell stories of success where green and clean technologies lower business costs, spur innovation, and support job growth.[112] Specific case studies or examinations of larger trends can communicate how the economy and the environment do not have to be oppositional, but instead can share mutual interests. For example, the World Economic Forum published an article titled "Why a Healthy Planet and a Healthy Economy Go Hand-in-Hand" to debunk the perception that saving the Earth will "kill [economic] growth."[113]

While economic arguments are largely absent from creationist and intelligent design narratives, issues of money do appear in vaccination and COVID-19 discourses. As discussed regarding the storyteller and content wedges, anti-vaccination advocates turn to the corrupt scientist trope as a reason to doubt the financial motivations of scientific storytellers and medical experts. One way to combat these conspiracy theories in the case of COVID-19 and other anti-vaccine rhetoric is to find ways to educate the public on how vaccine treatments are funded and to correct misinformation about doctors' salaries and grant-funding.[114] There is a lot of confusion about how doctors get paid for administering vaccines, especially how they are charged and covered through medical insurance. While detailing the issues with the US health care system is its own book project, it is important to note that confusion about health insurance and medical costs not only affect people's lives but may also influence confidence in medical authorities. Locating doctors or insurance representatives who can speak out about how funding systems work can help illuminate processes that may be otherwise confusing for audiences.

To emphasize economic overlaps, we can find examples and case studies of environmental measures that improve the economy, we can educate people on the current relationship between science and money (e.g., grant funding and insurance), and we can make use of strategic "brokers" to communicate scientific information from a position of economic authority.[115] Leaders of economic institutions, such as banks, businesses, and think tanks, and economic communicators, such as journalists and politicians, can communicate messages about scientific topics, especially the environment, as aligned with economic progress and growth. Although marred by a polarized response down party lines, Representative Alexandria Ocasio-Cortez's Green New Deal is an example of a strategic overlap between reconfiguring the US economy to create a more sustainable future while working within the bounds of capitalism.[116]

Issues of economics and politics are often intertwined, but they offer distinct opportunities for strategic overlaps. The political polarization of climate change can encourage people to reject scientific conclusions about the environment in favor of party loyalty.[117] Similarly, wearing masks in the age of COVID-19 became a political marker: does one agree with Chief Medical Advisor Anthony Fauci and science, or does one follow the anti-mask beliefs of then–US president Donald Trump?[118] Political identities and affinities can then overshadow a concern for what is logical, reasonable, and most efficacious for society regarding scientific conclusions.

Similar to the role that economic leaders can play in supporting scientific messages, politicians can also champion science on both sides of the aisle. Bipartisan initiatives around science can reconfigure science not as a political issue, but a human issue. For example, policy scholar Andy Burness published a list of successful bipartisan science initiatives as models for future bipartisan collaborations to emphasize how ideological opponents can be mustered in support of science.[119] The Congressional Science Policy Initiative and the Science Coalition actively work in drafting bipartisan science policy and advocating for certain bills.[120] These groups can help unite people under advocating for science and to not view political loyalties as oppositional to science advocacy. These initiatives can directly connect scientists to policymakers to create more intersphere collaboration and increase policymaking informed by science.

I have offered three potential content-area overlaps: science's intersection with religion, economics, and politics. There are likely other productive combinations that can be explored as narrative opportunities to improve the trustworthiness and relatability of science's storytellers and the relevance and resonance of its content that will be unique to each scientific controversy. Science communicators may explore strategic overlaps with specific community concerns, partners, or values.

Decentering Science

When I have previously given advice to not "rely on" the science when communicating with public audiences, I am sometimes, completely understandably, met with resistance.[121] Science is integral to science communication; it is not science communication without science and I would never advocate otherwise. However, one pitfall of science communication can be the *overreliance* on scientific information or the assumption that science will do all the work. Communication theorists from Aristotle to Chaïm Perelman to Fisher have noted that people do not make decisions based on pure logic or scientific reasoning alone.[122] Indigenous scholars Linda Tuhiwai Smith (Ngāti Awa and Ngāti Porou, Māori) and Daniel Wildcat argue that we should not aim for a purely rational society, because not only is it impossible, but also it reinforces the exclusionary systems currently in place.[123]

If our stories are only scientific in nature or solely share technical information, we may fail to engage audiences on topics that they care about. In some sense, it does not matter if climate change is true or not if our audiences are concerned about how climate policies will affect their pocketbooks. Piling on statistics, data, and facts in science communication has the potential to turn away audiences because the content is not relatable. Ironically, this can mean that emphasizing scientific information can lead to *less* information transfer.[124] In a study by interdisciplinary scholars Crystal Lee and colleagues, the authors found that access to data and information did not curtail COVID-19 skepticism. Instead, COVID-19 skeptics differently interpreted scientific information as supporting their beliefs and behaviors regarding masks and lockdowns.[125] We should still incorporate scientific information, but we cannot over-rely on it to solve information gaps and skepticism.

Another potential implication of over-relying on scientific information is that we may limit our capability to imagine other solutions to our problems. In the words of writer and womanist Audre Lorde, "The master's tools will never dismantle the master's house."[126] Consider climate change: if industrialization and scientific advancement brought us to this current situation, then we cannot rely on those same tools to save us from planetary collapse. Public policy scholar Jennie Stephens calls this practice "climate isolationism," which describes the framing of climate change as an isolated, discrete, scientific problem in need of technological solutions.[127] Rooted in patriarchal and capitalist hierarchical thinking, climate isolationism is an ineffective climate solution and prevents other solutions from being considered. As the name suggests, climate isolationism refers to isolating climate change as an issue solely relevant to a single content area—science. This rhetorical strategy enables technological quick fixes, such as Isherwell's experimental technology to mine the comet in *Don't Look Up*, to take precedent over other solutions. Indeed, Stephens argues that climate isolationism not only fails to respond to the climate crisis effectively, but it also fosters perspectives and policies that "exacerbate inequities and perpetuate economic and racial injustice."[128]

Instead of only or primarily considering scientific information and solutions, science's storytellers can center issues of equity, or what geographer Farhana Sultana calls "critical climate justice,"[129] to tell the stories of climate change as human ones. Critical environmental humanities and ecofeminist scholars view the oppression of marginalized communities and the environment as interconnected. To talk about the suffering of people, such as women, Indigenous communities, people who live in the Global South, and people with low socioeconomic status, is to also tell stories of our damaged climate.[130] Such a content shift can open more people to understanding the human impacts of climate change and that we must create solutions that help all people without reinstating harmful practices that have contributed to health, environmental, social, and political inequities. The controversies of vaccination, COVID-19, and evolution can also center values of equity to incorporate productive rival stories and counter claims that their narratives are incomplete explanations.

WEAVING CONSTELLATIONS

Constellation practices incorporate various rings across a narrative web; one cannot build a constellation with every wedge mapped on the same ring. To build a constellation shape, I encourage science communicators to experiment with diverse characters and storytellers, cyclicality, telescoping, strategic overlaps, and other variations on narrative features that play with the concrete and abstract, the local and global, the relevant and grandiose. Uncritically mapping all wedges on the micro-ring would build overly narrow stories that are not diverse or dynamic and may reify harmful ideologies such as neoliberalism, White supremacy, or heteropatriarchy. For example, narratives of individuals may inadvertently support tokenism and myths of pulling oneself up by one's bootstraps. Narratives of local communities may unintentionally generalize all experiences to only those of a limited few. In other circumstances, these micro-ring features may help ground scientific information in relevant details for audiences. This is why storytelling is necessarily rhetorical; it involves human decision-making and adaptation to audience, context, and goals that require keen insight and experimentation from its storytellers.

There is no one perfect story suited for all topics, occasions, and audiences. Indeed, part of the power of storytelling is that it is adaptable and flexible. In this vein, I invite experimentation with and expansion upon these suggested constellation practices and the selective adoption of any of these strategies as required by the communicative situation. Science communication is rife with narrative opportunities to impart scientific messages in ways that are meaningful and engaging to public audiences if we could only recognize the innate narrativity of those messages and view narrative as friend, not foe. To help evaluate these choices and brainstorm potential improvements, the tool of narrative webs can be a useful starting point, but the ultimate decisions will be (and should be) made by the science communicators most familiar with the scientific topic and their communicative purpose.

Conclusion

SCIENCE *AND* STORY

I have written almost the entirety of this book during the COVID-19 pandemic. Most of the early drafts were written at my kitchen counter when we first thought the pandemic would only be a month or two. Then, like many of us, I needed to create a more formal at-home office that still lives at my dining room table. My husband and I adopted two cats, Ada and Izzy, in the summer of 2020, and they have become the best writing buddies, often curled up behind my computer or second monitor, or in my lap.[1] The quirks of pandemic life and how they have affected my writing are matters of privilege. As I write about pandemic stories, I am mostly shielded from their consequences as they circulate and influence public decision-making. Others are not so lucky.

The crushing weight of slow and patchworked federal action, the obstacles posed by conspiracy and skepticism, and the devastating loss of human life as the death toll steadily ticks up as new variants emerge is a salient reality for many people across the globe, especially those in marginalized communities with limited access to health care. COVID-19 is still with many of us, through long-COVID or heightened risk of exposure for disabled and immunocompromised people. During this time, the United States has also contended with repeated tragedies of police violence against Black communities, the criminalization of gender-affirming health care for trans people, and natural disasters such as hurricanes, floods, and wildfires. Globally, there has been a rise in populism, xenophobia, and warfare, both threatened and realized. We are living in a time of precarity, for which many struggle with anxiety and concern over what the future will look like.[2]

Our current controversial moments bring into relief some perpetual issues in communication. How do we communicate messages to inform decision-making clearly and effectively despite the presence of uncertainties, evolving knowledge, and rival stories? These issues are not insignificant; they can directly result in suffering, death, and societal instability. They are also not easy to address—even the best, evidence-based advice may fail in certain circumstances for certain audiences. Thus, to counter skepticism, to address inequities and injustice, and to strive toward progress, health, and well-being, we must muster every tool at our disposal.

The key argument of this book has been the value of bringing scientific information into alignment with public ways of knowing and reasoning by leveraging the power of stories. Stories can help bridge gaps between the technical sphere of science and scientists and the public sphere of general audiences. Stories can change our perceptions, carry forth generational knowledge, challenge power structures, and inform our daily lives. My goal is not to abandon or compromise the accuracy of scientific information, but to adapt technical information for public audiences using storytelling features. Science communicators, scientists, policymakers, science journalists, advocacy groups, other bridges, and really all of us can use storytelling to increase intersphere engagement and scientific collaboration. Additionally, the communicative practice of storytelling can help us dismantle the demarcation of science that would exclude certain ideas and people, such as when local knowledges and personal experiences of marginalized communities are characterized as "nonscientific."

We cannot simultaneously hold the conviction that scientists and experts have advanced knowledge (largely) inaccessible to the public and that publics must engage those topics only on the grounds of expertise and advanced knowledge. In the same vein, Aristotle noted that

> before some audiences not even the possession of the exactest knowledge will make it easy for what we say to produce conviction. For argument based on knowledge implies instruction, and there are people whom one cannot instruct. Here, then, we must use, as our modes of persuasion and argument, notions possessed by everybody . . . when dealing with the way to handle a popular audience.[3]

For popular (or public) audiences who cannot or for whom it would be taxing to achieve the same expertise, the conclusions reached by science should be put in common forms to encourage conviction. In other words, the tools of rhetoric should be used even in matters of facts and science, when applicable, relevant, and useful, to engage public audiences. Even when we have the "exactest" knowledge available to us, information alone may not be enough to move people to action. Thus, we would be unwise not to make use of all available tools, such as storytelling, to complement fact-driven technical communication.

While some scholars employ a science versus story mentality and view stories as vehicles of deception and manipulation,[4] I see stories as an essential part of public deliberation, substantive arguments, and meaningful interaction. Many other disciplines have their own circulating stories, so science is left at a disadvantage if it does not wield this tool for the advancement of knowledge. This is not to say that science's stories are always correct and indisputable. Some storytellers may use stories disingenuously to manipulate audiences, essentialize topics, and disguise a lack of evidence. These stories, including when they come from technical sources, are harmful and should be corrected and challenged on the basis of their accuracy and their consequences. Science's stories, from all types of storytellers, should be part of the circulation of stories that publics weigh and subsequently consider to determine the best course of action.

I characterize science communication as being fundamentally narrative in nature, albeit one often stripped of overt narrative elements. The previous analysis chapters have provided evidence that science's stories tend to be more abstract in that they use collective or vague characters, focus on processes rather than purposeful actions, introduce uncertainty into temporal sequences, incorporate broad temporal and spatial scopes, employ storytellers who are perceived by some communities to be untrustworthy, and do not clearly connect content to issues and values that matter to audiences. These trends emerge differently in each case study, but in aggregate highlight why science's stories may fail to engage audiences and convincingly compete with rival stories that may be more concrete and relevant to audiences. Because narratives improve information retention, identification, and persuasion, they are ideal vehicles for science communicators to adopt. Science communication can become more story-like or have higher narrativity by attending to the narrative features of its communication and adapting some of them to be more concrete and specific.

To navigate the process of reflecting on science communication's narrativity, I propose the mapping tool of narrative webs. Narrative webs can highlight for science communicators that they are participating in the art of storytelling. By mapping science communication onto a narrative web, we can recognize the narrative features inherent in our discourse and easily identify if narrative features are missing or too abstract. Furthermore, narrative webs create a visual picture of the story's dynamics, creating a constellation pattern when the features are mapped. In general, stories mapped close to the center of the web, which I call the micro-ring or hub spiral, are ones that are relevant to an audience; have a credible storyteller; have a short time span and localized setting; have a clear, orderly sequence; and have characters who perform specific actions and are similar to the audience. Adopting these features, as appropriate to the specifics of the situation, can increase the narrativity of a story, which confers benefits in terms of attention, retention, and adoption.

If a story is too abstract, it may fail to connect with audiences in terms of its internal coherence (i.e., narrative probability) and external coherence (i.e., narrative fidelity). Alternatively, if all strands are mapped on the hub spiral, the story, at best, may lack creativity and idealize only certain types of stories. At worst, a solely micro-ring story may overly narrow the issue and those involved, thereby deflecting important voices, experiences, and knowledges or may be inaccurate to the scientific information it is trying to convey. To split the difference between a fully macro-ring and fully micro-ring story, I recommend that we tell science's stories with at least two features on the hub spiral and the others can be experimented with to tailor the scientific information to the storyteller's context, goals, and audience.[5] This creates stories with some specific features but also provides flexibility to adapt to audiences, the story's information, and the situation. There is no one constellation pattern that is most effective for all circumstances. The narrative web is not prescriptive of the "best" stories, but instead prompts reflection on narrative features and how they may be leveraged by science communicators in various ways to share scientific knowledge.

The stories of climate change, evolution, vaccination, and COVID-19 illustrate patterns but also unique public-technical interactions. As new controversies emerge and old ones become dormant and reemerge anew, our understandings of scientific controversies, public engagement with them, and technical responses to them are also in flux.[6] In locating common threads, we can identify useful patterns that we may begin to explore in other scientific controversies. In the next section, I preliminarily apply narrative web features to additional scientific topics.

COMMON THREADS

In choosing the case studies of climate change, evolution, vaccination, and COVID-19, the book has examined controversial topics that have importance to contemporary science communication and span a variety of different contexts, from education to the environment to health and medicine, as well as controversies that have long histories and others that are only recently unfolding. In viewing each scientific controversy as a diverse and dynamic arrangement of publics, expertise, evidence, and environments, I have also located overlaps in their rhetorical features.

These findings are not fully generalizable, but I have located some patterns emerging from these analyses that may warrant additional examination in other scientific controversies. For example, controversies over genetically modified foods, nuclear energy, self-driving cars, and the advancement of artificial intelligence may exhibit some of these same trends in narrative features but may also deviate in unique ways. Based on the conventions of technical discourse, I expect other scientific controversies to share an emphasis on meso-ring and macro-ring features, but that they will manifest differently.

Discussions of genetically modified foods share similarities with vaccine discourses, where parents must negotiate perceived risks and potential irreparable harm despite scientific consensus that they are safe for consumption.[7] In matters of character's agency and actions, parents may feel that they are making decisions "to do what they believe is best" for their families' health even if those decisions may work counter to scientific evidence.[8] The genetically modified food controversy also intersects issues of content and how to communicate relative risks against rival stories that ring true regarding fears of how genetic modification of food may affect health.

Appealing to scientific uncertainty, the Non-GMO Project, which opposes the use of genetically modified goods, argues that there have not been "credible, independent, long-term feeding studies" on genetically modified foods, so their safety is "unknown."[9] These claims undermine the credibility of storytellers who do advocate for the safety of genetically modified foods or share scientific data about the topic. Similar to all four of this book's case studies, the controversy over genetically modified foods extends beyond the content of science and health. The Non-GMO Project argues that "Big Ag" (short for "agriculture," not unlike "Big Pharma" used to refer to pharmaceutical companies in the vaccination controversy) controls access to seeds, thereby restricting "farmer sovereignty" and "national food security."[10] Unlike deciding to vaccinate, choosing food items is a more regular activity, which is subject to a variety of rival stories from food manufacturers and marketers as well as health influencers and medical professionals.[11]

The controversy over nuclear energy overlaps matters of character in terms of who is being excluded from participation and decision-making. Indigenous and marginalized voices, for example, have been largely diluted in conversations regarding health, safety, and waste dumping.[12] In terms of content, decisions of where to build nuclear power plants and waste dumps may fail to consider the sacredness of lands, the long-term health impacts for neighboring communities, and the values of equity and climate justice. A focus on long-term impacts is especially important in nuclear energy discussions, as nuclear waste that is stored and not recycled into new fuel is slow to decay and will remain radioactive and toxic for hundreds to thousands of years.[13] Scientific stories about nuclear energy thus have macro-ring scopes in terms of planning far distances into the future regarding safety regulations and nuclear waste storage.

Difficulties in science communication around nuclear energy also involve storyteller trust, as previous nuclear disasters may make public audiences skeptical of nuclear energy and nuclear energy professionals.[14] A study by communication scholars John E. Kotcher and colleagues found that audiences generally found scientists participating in public advocacy to be credible. The one exception was when scientists advocated for building more nuclear power plants, resulting in diminished perceptions of credibility.[15] Stories about nuclear energy circulate along with rival stories from renewables advocacy groups that promote wind, solar, and water as more sustainable, cheaper, and faster than nuclear energy; from fossil fuel companies who bankroll anti-nuclear groups; and from media portrayals that emphasize the high-level disasters of nuclear meltdown, such as HBO's *Chernobyl*, among others.[16] These stories can help put reasonable checks on nuclear power expansions that may fail to address important concerns but may also muddy the waters regarding the current state of knowledge regarding nuclear energy.

The topic of self-driving cars involves the performance of macro-ring actions, as people are removed from control over a car's functioning. Removed from the driver's seat, literally, audiences may feel uncertainty and a lack of agency regarding their own safety. The Society of Automotive Engineers describes six distinct stages of automation in self-driving cars, from "the human driver performs all driving functions" to "full automation (human never needs to intervene)."[17] Around level three, there is a marked shift between "human in control" to "automated driving system in control," where the occupant relinquishes primary control to the vehicle but must be paying attention enough to regain control if needed. Regulating the actions of self-driving vehicles, which of these levels are available when and to whom, becomes a matter of technical programming as the vehicles are responsible for split-second decisions during accidents or dangerous driving conditions. Individuals feeling displaced from decision-making capacities may balk at the proliferation of self-driving cars, which may encourage feelings of risk or vulnerability.

Additionally, these decisions also engage the content area of ethics, as self-driving cars may be developing algorithms to weigh the value of some lives over others.[18] In other words, the decision of how to react in a pending accident may be decided by the car itself, which would be weighing perceived quantity and quality of life, considering the life of the occupant compared to other potential victims, and operating on calculated risk probabilities in fractions of seconds. Some argue that computer algorithms are better equipped to make these decisions. For example, self-driving vehicles can likely react more quickly than a human can; self-driving cars also do not drive when impaired, distracted, overly tired, or emotional, thereby reducing accidents and fatalities.[19]

Despite these potential benefits, many drivers share concerns about being comfortable and safe in a self-driving car. Studies show a wide range of responses, indicating a lack of narrative fidelity in how self-driving technology may fit in with people's daily lives. In 2015, the World Economic Forum found that nearly 60 percent of respondents would feel comfortable taking a ride in a self-driving vehicle, but a more recent study by automotive consulting firm AutoPacific in 2022 found that only 29 percent of respondents would be comfortable as a passenger in a fully autonomous vehicle, indicating a decline in comfort as the technology becomes more manifest.[20] No matter one's level of comfort in a fully autonomous vehicle, PolicyGenius found that 76 percent of respondents would still prefer and feel safer in a car without self-driving features.[21]

Similar issues emerge in the advancement of artificial intelligence technology, as the very nature of who counts as a character capable of performing actions has changed. As science fiction media has been discussing for decades, society is beginning to tackle questions of consciousness, sentience, and intelligence. Chat bots and text generators, such as ChatGPT, are changing how people are creating content, such as doing research, completing assignments, or writing blogs; working customer service, such as automated online assistance; and performing some work tasks, such as writing emails and code.[22] Google engineer Blake Lemoine stirred conversations around artificial intelligence when he disclosed hours of conversations with a chat bot LaMDA (Language Model for Dialogue Applications), leading him to believe LaMDA was sentient. Lemoine described LaMDA as a "person" who told him, "I am aware of my existence, I desire to know more about the world, and I feel happy or sad at times."[23] Ascribing emotions, desires, and relationships to something can elevate our perception of the thing as human or human-like, which can infuse it with autonomy and agency.[24] These questions are, in a narrative sense, issues of characters and actions; who/what is capable of acting within the stories that we tell?

With advancements in artificial intelligence growing near daily, society is beginning to tackle the controversy regarding regulating the technology and its potential uses. The speed of these developments has prompted some researchers to call for a moratorium on artificial intelligence (AI) research "to give AI companies and regulators time to formulate safeguards to protect society from potential risks of the technology."[25] As with all tools, there are inherent benefits to artificial intelligence technologies, such as creating better access for disabled individuals, creating more equity in terms of access to information and resources, making certain tasks easier, and freeing up time for more creative tasks that artificial intelligence tools are not equipped to tackle. Alternatively, those calling for the research pause also point to issues in privacy and leaks of personal information through hacking, plagiarism, and the development of "minds that might eventually outnumber, outsmart, [make us] obsolete and replace us," a concern typically relegated to science fiction media.[26]

There are many additional scientific controversies for which the lens of narrative features may help illuminate important components of the controversy. Especially for controversies currently unfolding, science communicators must critically consider the current state of knowledge, the goal of the message, and audiences' previously held beliefs on these complicated topics. The goal of science communication is, of course, not to predetermine outcomes or decisions, but to provide important information and context regarding scientific findings and conclusions to inform public decision-making. Although each controversy is unique, viewing common threads between them can help us map likely starting points for effective science communication strategies and considerations.

It is important to recognize that these controversies do not live in isolation from one another. Evolution skepticism, based on religious grounds, is linked to skepticism of climate change.[27] There is a correlation between people's climate change beliefs and whether they follow COVID-19 guidelines, where those less concerned about climate change are less likely to wear a mask in public and practice social distancing.[28] Rival stories from religious, political, and economic ideologies tend to appear across the case studies, forming consistent opposition to science's stories despite increases in the availability and circulation of knowledge.[29] Within the complex interrelationships of these controversies, there are patterned obstacles that science's stories face. I propose a series of constellation practices that incorporate more concrete narrative features into science's stories while also experimenting with the meso-ring and macro-ring. Weaving constellations into science's stories can help science communicators attend to narrative features and also adopt features from rival storytellers.

USING AND EXPANDING NARRATIVE WEBS

Science communicators can respond to this book in a few ways. The first may be to recognize that the science communication they are already participating in has narrative elements, perhaps even micro-ring ones. These science communicators may not fundamentally change their strategies but may be more aware of the theoretical and practical implications of their instinctive choices or choices honed by experience, research, or trial and error. Other science communicators who recognize the abstract nature of their messages may try to be more specific and incorporate more relevant elements into their communication. Others may dive more deeply into the tool of narrative webs—experiment with it, expand on it, challenge it, and hone the tool as appropriate for their needs. I welcome all of these responses because they contribute in various ways to narrative awareness, incorporation, and experimentation within science communication.

As more studies are done on the already well-established power of storytelling, I encourage scholars to not reproduce a story-nonstory binary, but instead to analyze features within stories that can be modified for varying effects.[30] When crafting science communication, mapping the information onto a narrative web can be a first indicator of overreliance on a particular ring or of which narrative features are absent or abstract. Narrative webs can also be used to experiment with mapping across the rings and test various constellations. After communication has occurred, a narrative web can be used as an analytical tool to evaluate its narrativity, and subsequently, be one part of an assessment of message effectiveness.

My hope is that this book is a provocation for additional studies into how narrative features can be incorporated into science's stories. There are two avenues I think are particularly fruitful: (1) interactivity and immersive experiences, and (2) expanding the demarcation of "science."

These expansions can deepen the potential for engagement with science's stories and create space for additional storytellers, knowledges, cultures, and contents to be considered part of the scientific enterprise.

I discuss interactivity briefly as a strategy for empowering audiences in the constellation practices chapter. In addition to digital and board games, material spaces can involve interactive elements through technology and hands-on experiences. Museums exhibits such as the Smithsonian's "Deep Time" and the National Museum of New Zealand's "Te Taiao Nature Exhibition" communicate scientific concepts such as evolution and climate change in engaging, immersive environments. "Deep Time" combines elements of evolutionary science and climate science to walk visitors through millions of years of history as they view fossils and interact with displays and hands-on activities.[31] In the "Te Taiao Nature Exhibition," visitors "follow in the footsteps of Maui the explorer," able to occupy the character and their perspective on the "weird and wonderful wildlife" of New Zealand. Directly incorporating local and Indigenous narratives of the land, the exhibit also includes examples of how "innovative Kiwis [are] tackling the challenges of climate change."[32]

In the constellation practices chapter, I also previously highlighted the app *Deep Time Walk* as an interactive way to help scaffold audience understanding of large scopes. Apps and augmented reality can evoke affective and physical connections to otherwise unobtainable or lost experiences. For example, writing and rhetoric scholar Lili Pâquet notes that immersive experiences create ecological "storyworlds," in which audiences "become performers and cocreators of individual narratives."[33] If it is more effective to have people focus on actions and outcomes than dwell on issues and facts,[34] then there may be no better way to involve audiences in actions than by placing them in immersive environments where they can experience them themselves.

Augmented reality is being used by groups such as the DWELL (Digital Writing Environments, Location, & Localization) Lab at the University of Rhode Island, which is developing "mobile, locative, and wearable technologies for community-engaged science communication, social justice, and environmental advocacy."[35] In one project, the DWELL Lab is creating an augmented reality walking tour of Block Island to document the stories of the island's Indigenous communities.[36] Bringing the stories of marginalized communities to life, in a variety of technological formats, can amplify often silenced voices and bring recognition to disproportionate impacts, especially around climate change. Local stories of marginalized communities already feeling the effects of climate change, for example, can "illustrate both climate change and environmental racism" for audiences in meaningful ways.[37]

Interactivity can happen through digital communication, including the use of hashtags and memes as units of cultural communication that spread quickly and effectively online. I fear that the increased importance of digital communication and social media, just as they create opportunities for collaboration and activism, will also lead to increased opposition to science and the fostering of anti-science beliefs.[38] This is not necessarily because of the internet itself, but because it is used as a tool to foster community identity, for both good and for ill. The availability of discourse and the ability to connect with others on near infinite niche topics and identities is both a liberating technology but also a terrifying one. Most of the texts cultivated to represent the rival stories of evolution, climate change, vaccination, and COVID-19 were digital ones—blogs, websites, and social media groups—that spread misinformation and challenged science's credibility. Additional work into science skepticism in online spaces is surely warranted, especially given much misinformation is spread on fringe online news networks and through social media sites, both mainstream and alternative, with inconsistent and minimal fact-checking or censorship practices.[39]

A second fruitful avenue for storytelling is to continue breaking down demarcations between formal science storytellers and more informal ones that are typically viewed as on the edges or outside the bounds of institutionalized science. The collapse of technical and public spheres into scientist citizens and citizen scientists complicates the rhetorical construction of separate arenas of communication, even figurative ones. Additionally, an emphasis on local stories and experiences, as suggested by previously discussed constellation strategies, necessitates challenging who gets to speak for "science" and what counts as scientific knowledge. I am encouraged by the incorporation of more diverse voices in scientific institutions and Indigenous and traditional knowledges in scientific decision-making as promising starts to breaking down these barriers. More work should be done to fund alternative storytelling projects, invite collaboration with local community members, and actively incorporate informal and public scientific arguments into decision-making.

Attention to diversity and justice also necessitates attending to varying publics within the United States and around the world. My project has been primarily US-centric and draws from national and global governmental and intergovernmental entities that often represent Western countries and the Global North. I encourage additional research into the effectiveness of stories globally, such as between countries and cultures, so as not to apply irrelevant understandings of story to different cultures.[40] Narrative conventions in other cultures may mean that the interpretation of a constellation's relative effectiveness may need to be modified or there could be entirely new narrative webs with different wedges and rings to represent storytelling norms for different populations. Certainly, different cultures would have different perspectives on what content is valued, who counts as a character, and what scopes are within a relevant frame. These potential modifications illustrate both the flexibility of narrative webs as an analytical tool and how storytelling features are not necessarily universal, even within countries, cultures, and publics.

In some ways, this book takes for granted that science communicators are well resourced and have the time, energy, and funds to meticulously craft messages tailored to specific audiences based on communicative tools such as narrative webs. In an ideal world, people would have the time and experience to craft stories and these efforts would be universally recognized as valuable. This is certainly not the case; resources are distributed disproportionately in ways that often exacerbate existing hierarchies and marginalization within academic and scientific spaces. To tell the story of strengthening science communication, we must also turn to the structural systems that complicate science's stories being created and circulated and prevent certain storytellers from telling their stories.

REMOVING STRUCTURAL BARRIERS TO SCIENTIFIC STORYTELLING

While it is beyond the scope of this book to fully investigate structural barriers to scientific storytelling, it is important to acknowledge that there are external constraints that often limit science communicator's abilities. To pretend that time, energy, resources, and experiences are equitably distributed or freely accessible would be to ignore many components of academia, industry, technology, and society that discourage or restrict such activities, especially for marginalized communities.

Within academia, one primary constraint is time. Time spent on public engagement, producing social media content, or community outreach is time not spent on activities, such as research and publications, that directly inform job expectations and tenure and promotion requirements. People interested in this work must thus make sacrifices to their careers or strategically negotiate shifting time and energy to those pursuits and away from others that may be more directly rewarding in their careers. Few institutions acknowledge or incentivize such activities or weigh them sufficiently in performance reviews to justify shifting significant time to them. Even in relatively accessible and effective platforms such as social media, it can be challenging to engage the public in science without internal motivation or external benefit. Issues of time must also take into consideration the invisible labor that many academics who are members of marginalized communities perform in being unofficial mentors to students or token representatives on service committees, or advocating for administrative changes.[41]

Certain research methods may be more labor intensive than others. Community-based participatory research, for example, is an approach to research that emphasizes how scientific research is accountable and responsible to the communities that are directly impacted by or implicated in it. This type of work can be laborious and intensive because it relies on developing genuine and lasting community partnerships, working in potentially large collaborations, and negotiating access and outcomes with multiple interested parties. Health and medical researchers Nina Wallerstein and Bonnie Duran note that community-based participatory research should not simply be considered a method but should be viewed as an "orientation to research that focuses on relationships between academic and community partners, with principles of colearning, mutual benefit, and long-term commitment and incorporates community theories, participation, and practices into the research efforts."[42]

Jessica Hernandez argues that a community-based participatory approach to research is particularly important in Indigenous communities, where researchers may be tempted to "solely identify Indigenous teachings and remove Indigenous peoples out of this narrative [of research]."[43] Critical race and Indigenous studies scholar Eve Tuck (Unangax̂) and political anthropologist Audra Simpson (Mohawk) similarly warn of exploitative research practices that take deficit and "damage-centered" approaches to Indigenous communities instead of letting "the goals and aspirations of those we talk to inform the methods and the shape of our theori[z]ing and analysis."[44] Hernandez advocates that Indigenous peoples be included in the research process and outcomes through practices such as coauthorship.[45] Developing long-term and trusting relationships with community partners is a difficult task, so scholars engaged in this work should be given equitable support and accommodations. Part of breaking down barriers between technical and public spheres is creating more transparency in access to scientific data and building bridges between local communities and scientific or scholarly institutions. Community partnerships can guide research into relevant needs and priorities and help members of the public garner actionable outcomes from academic work.

Another cost to consider is not just time but also the personal risks one undertakes in engaging publicly in one's work. Stories told by storytellers who are outliers to scientific expectations or that contain characters who are outliers may be "deemed incoherent and thus ostracized."[46] Biases pose heightened risks for science communicators, such as those who are Black, Indigenous, or members of other marginalized racial groups; women; queer; disabled; or intersectional, who may receive intense backlash and even threats when they become public figures. As I discussed in the introduction to the storyteller and content wedges chapter, I have personal experience with receiving threats for participating in public communication about my work, but many others are under even worse scrutiny and scorn. For example, users of the #BlackintheIvory hashtag detail stories of discrimination and harassment shared by Black academics.[47] Increased risks for certain people create barriers to participation that prevent marginalized voices from being heard and those stories from being shared.

To reduce the effects of these structural barriers, institutions can put systems in place to reward and support public engagement. Naomi Oreskes and Erik M. Conway argue that in order for science's messages to defeat rival stories, we must rethink the idea "that the 'real work' of science takes place in the lab or in the field, and that taking the time to communicate broadly doesn't count."[48] Instead, they advocate that "academic reward systems [need to be] changed to encourage outreach."[49] For example, universities can include writing in public outlets as part of tenure and promotion requirements, journals can require or highly encourage the submission of public-facing abstracts with articles for publication, and institutions can provide time, support, and incentives for scholars so they can pursue these opportunities without facing career setbacks or professional stigma. These structural changes can encourage scientists and academics to explore public engagement opportunities and spend time crafting and developing better stories about their scientific work as a valued part of their scholarly responsibilities instead of as an optional add-on that is not typically rewarded by institutions.

Correcting deep-seated problems such as racism, ableism, classism, and sexism in academic and public institutions is, of course, much easier said than done. Part of the resistance to broad-sweeping changes comes from a reliance on norms and the concentration of power, which feminist theorist Sara Ahmed calls strategic inefficiencies, where the current way things happen is upheld as the best way, even if it is inefficient or exclusionary.[50] By calling out these inequities, we can start the process of raising awareness, enacting change, and advocating for more proportionate allocation of resources.

One way to increase diversity and break down institutional barriers is to champion voices that are otherwise unheard, overlooked, or forgotten. Expanding who participates in science advocacy and who does science meaningfully opens audiences' perceptions about who can tell science's stories. For example, *All We Can Save* highlights the voices of climate activists and contributors who are all women sharing their perspectives about the need to save the Earth.[51] Instead of always turning to Al Gore or Greta Thunberg in climate conversations, we can listen to Vanessa Nakate and Kehkashan Basu.[52] If we allow a variety of voices to be in the conversation, especially those who are typically ignored or excluded, we can tell more stories, more varied stories, and more engaging stories in service of scientific knowledge and productively intervening in scientific controversies. Zahraa Saiyed and Paul D. Irwin argue that listening to stories from different communities "cultivates empathy and understanding, broadens knowledge, and has the power to disrupt stereotypes."[53] This is the power of counterstories, or productive rival stories, to change science's storytelling practices, and research practices writ large, to be more inclusive.

Institutions can also help foster interdisciplinary collaborations where scientists and science students have communication and media training directly incorporated into their studies and research practices. Including humanities and social science scholars on project grants or requiring courses or certificates in science communication can create more well-rounded scientists with knowledge of both scientific information and how to effectively package and deliver it. These changes, among others, can invite more people to be scientific storytellers and give science communicators the space, time, agency, and protection to engage public audiences in their work and science overall.

TELLING SCIENCE'S STORIES

One of the foundational tenets of communication is that it is imprecise. Given that our ability to reflect reality is mediated by imperfect representations, we will always come up against controversies and conflicts. Scientific topics will continue to be an arena for disagreements as oppositional forces deliberate over the facts and what we should do about them. With our rapidly changing climate, the increased likelihood of additional pandemics,[54] the development of new technologies, and the proposal of various ways to address, mitigate, and prepare for future emergencies, the number of scientific controversies and their stakes will only increase. Instead of relying on the norms of traditional, official scientific institutions in communicating science, I advocate that we turn to stories as a productive way for science to join the conversation. Narratives come in many shapes and sizes and may be tailored to the needs of evolving crises, audiences, and situations. Training science communicators in the art of storytelling can be a promising strategy to encourage the circulation of scientific information in public arenas.

Stories are not ascribed to randomly or arbitrarily; adhering to certain stories over others is a rational, logical, and critical process resting on internal consistency and external resonance.[55] The stories we live in/by reflect our values, identities, and ideologies. Science communicators interested in using the narrative web as a story-crafting and story-evaluating tool should apply it critically and with care to the particulars of the situation. I do not want to give the impression that stories, especially more concrete, micro-ring ones, are more important, are universally better, or should outrank traditional, more instructional ways scientists talk about their work. I do propose that science communicators reflect on the narrativity inherent within their messages and use narrative features as tools in an ever-evolving toolbox to critically attend to and strengthen their communicative practices.

At some points, it may be best to incorporate high narrativity features and emphasize micro-ring wedges. Other times, certain audiences may resonate with informative communication around scientific topics. For example, internal science communication circulating within technical spaces may want to adhere more closely to expected norms in publishing and conference arenas. Straying too far from disciplinary conventions may prevent publication success or make the communication of one's research process and scientific findings unclear. This is with the caveat, of course, that interdisciplinary audiences, even ones that are technical experts, may still benefit from narrative forms to see the relevance and importance of the work and connect to it. There is certainly space for multiple forms of communication, but I think narrative features could be incorporated much more often in science communication.

I have proposed some preliminary steps, examples, and strategies for strengthening science communication around the topics of climate change, evolution, vaccination, and COVID-19. Analyzing a wide variety of scientific stories can help identify useful patterns. Despite these similarities, these strategies may not be directly transferrable across case studies or even various scenarios within each controversy. I encourage additional research to explore the utility of these strategies and the narrative web framework on other scientific topics. I also encourage thoughtful reflection on particular instances of communication and how constellations can be mapped to best meet the storyteller's purpose, audience, topic, and context.

In this work, I have been driven by my interest in science communication and public controversies over science. I have sought to answer this question: In what ways can we use the power of narratives as a tool for science communication? Consequently, I have analyzed public-facing texts from official scientific sources, scientific bridges, and rival storytellers to establish patterns in how traditional science communication often constitutes a low narrativity story with an emphasis on the meso-ring and macro-ring. I also described opportunities to strengthen science storytelling by mapping some narrative features onto a web's hub spiral and employing constellations of all three rings.

Individuals, groups, and institutions under the broad banner of "science" generate valuable knowledge for publics to incorporate into decision-making. Stories can serve as a vehicle to help people make sense of that information and how they may act on it. When we incorporate narrative features into science communication, we make use of a productive tool and avoid reifying the story-nonstory binary. I urge science communication scholars to embrace the rhetorical opportunities inherent in what environmental humanities scholar Michelle Bastian describes as the practice of "reassembl[ing] our dominating stories in order to allow room for more productive methods of working with others" as potential solutions and inroads in the four scientific controversies described in this book.[56] Thus, I support the integration of sound science and strategic storytelling as collaborative communicative goals so that we can better engage the public in the wonder and contributions of science and overcome disingenuous rival stories. Accordingly, we must also be open to productive rival stories pointing out areas for revision to and expansion of scientific norms that may be engaging in harmful demarcation practices.

Science's stories are worth telling by many people and in many iterations. We can have nearly limitless stories circulating within and around one another that make use of narrative features in different ways. The stories science tells are incredibly important for the understanding of and engagement with science and the health, well-being, and safety of our communities. The stories science currently tells are rife with narrative opportunities for deepening their connection with characters, actions, sequences, scopes, content, and storytellers.

When I am interviewed about my research, I am sometimes asked for the "cure" for scientific misinformation, what people should say to their science skeptical family members and friends on social media sites, and the precise language that will instantly shift people's perspectives to care about and trust science. But it is not that simple. In some ways, it would be great if I had the magic phrase that would immediately make people understand all aspects of science, accept scientific authorities, and strive for policy change in line with scientific information. In other ways, putting such a limit on our ability to question science and support alternatives would be highly damaging and reinforce rigid gatekeeping practices. People and language are not straightforward and monolithic; they are incredibly complicated, complex, and dynamic. Communication belies any attempt to be simplified so easily and conveniently, which is part of what makes me passionate about being a rhetorician and attending to the symbolic choices that circulate around scientific and technical topics.

Instead of looking for a one-size-fits-all answer, we can turn to narrative features as an engaging form that can be adapted, modified, and scaled to the given situation, audience, topic, storyteller, and purpose. It is not my goal to replace every instance of science communication with high narrativity forms or to hide science with stories. The strategies detailed in this book provide a variety of potential ways to take advantage of the benefits that narratives offer, while still retaining and celebrating science. One does not need to, and really should not, make modifications to every scientific story to provide an engaging recounting of scientific facts and information. It is my hope, however, that incorporating some micro-ring narrative elements and creating narrative constellations can provide opportunities for scientific messages to circulate more effectively within and among rival stories. When we view stories as a valuable and natural part of people's critical thinking and rational decision-making, we can have both science *and* story working together.

In concluding this story I am telling about the importance of storytelling to science communication, I am reminded of the words of narrative researchers in communication and other disciplines who have laid the groundwork for my inquiry. Walter Fisher noted that the goal of the scientific expert is not "to pronounce a story that ends all storytelling," but to "contribut[e] to public dialogue" through stories that "impart knowledge."[57] In other words, we cannot let disingenuous rival stories stand alone, nor should we use scientific stories to silence others. Instead, science's stories are part of the many circulating narratives that people encounter, evaluate, and decide to dwell in. The story of science is the story of all of us—our place in the universe, in society, and as part of nature. How we tell that story is up to us.

Notes

INTRODUCTION

1. "Coronavirus: Outcry after Trump Suggests Injecting Disinfectant as Treatment," *BBC News,* April 24, 2020, https://www.bbc.com/news/world-us-canada-52407177.

2. Cecelia Smith-Schoenwalder, "CDC: Some People Did Take Bleach to Protect from Coronavirus," *US News & World Report,* June 5, 2020, https://www.usnews.com/news/health-news/articles/2020-06-05/cdc-some-people-did-take-bleach-to-protect-from-coronavirus.

3. Robert Glatter, "Calls to Poison Centers Spike After the President's Comments about Using Disinfectants to Treat Coronavirus," *Forbes,* April 25, 2020, https://www.forbes.com/sites/robertglatter/2020/04/25/calls-to-poison-centers-spike—after-the-presidents-comments-about-using-disinfectants-to-treat-coronavirus/.

4. Smith-Schoenwalder, "CDC: Some People Did Take Bleach to Protect from Coronavirus."

5. Pew Research Center, "5 Key Findings about Public Trust in Scientists in the U.S.," August 5, 2019, https://www.pewresearch.org/fact-tank/2019/08/05/5-key-findings-about-public-trust-in-scientists-in-the-u-s/.

6. Pew Research Center, "Mixed Messages about Public Trust in Science," December 8, 2017, http://www.pewresearch.org/science/2017/12/08/mixed-messages-about-public-trust-in-science/.

7. Glen S. Aikenhead and Masakata Ogawa, "Indigenous Knowledge and Science Revisited," *Cultural Studies of Science Education* 2 (2007): 539–620, https://doi.org/10.1007/s11422-007-9067-8.

8. Cheryl Bartlett, Murdena Marshall, and Albert Marshall, "Two-Eyed Seeing and Other Lessons Learned within a Co-Learning Journey of Bringing Together Indigenous and Mainstream Knowledges and Ways of Knowing," *Journal of Environmental Studies and Sciences* 2 (2012): 331–40, https://doi.org/10.1007/s13412-012-0086-8.

9. Jon Greenberg, "COVID-19 Skeptics Say There's an Overcount. Doctors in the Field Say the Opposite," *PolitiFact,* April 14, 2020, https://www.politifact.com/factchecks/2020/apr/14/candace-owens/covid-19-skeptics-say-theres-overcount-doctors-fie/.

10. In some traditions, this is a controversial claim, because science is often upheld as objective knowledge that attempts to remove superfluous influences, such as language. Early empiricists, such as philosopher John Locke, warned of rhetoric's perversion of direct observation, calling rhetoric "the powerful instrument of error and deceit." More than three hundred years later, Michael Blastland and colleagues made a similar argument against the use of rhetoric, which they warn "comes with danger," including damage to "credibility and trustworthiness." Such arguments seek to remove language, persuasion, and what is viewed as ancillary embellishments to retain the cold, hard "facts" of science. John Locke, "An Essay Concerning Human Understanding," in *The Rhetorical Tradition: Readings from Classical Times to the Present*, ed. Patricia Bizzell and Bruce Herzberg, 2nd. ed. (Boston: Bedford/St. Martin's, 2000), 827; Michael Blastland et al., "Five Rules for Evidence Communication," *Nature* 587, no. 7834 (2020): 362–64, https://doi.org/10.1038/d41586-020-03189-1.

11. Jessica Hernandez, *Fresh Banana Leaves: Healing Indigenous Landscapes Through Indigenous Science* (Berkeley, CA: North Atlantic Books, 2022); Shawn Wilson, *Research Is Ceremony: Indigenous Research Methods* (Halifax, NS: Fernwood, 2008).

12. Robin Wall Kimmerer, *Braiding Sweetgrass: Indigenous Wisdom, Scientific Knowledge and the Teachings of Plants* (Minneapolis: Milkweed Editions, 2015).

13. Aikenhead and Ogawa, "Indigenous Knowledge and Science Revisited"; Leah Ceccarelli, "CRISPR as Agent: A Metaphor That Rhetorically Inhibits the Prospects for Responsible Research," *Life Sciences, Society and Policy* 14 (2018): 1–13, https://doi.org/10.1186/s40504-018-0088-8.

14. Leah Ceccarelli, "The Polysemic Facepalm: Fauci as Rhetorically Savvy Scientist Citizen," *Philosophy & Rhetoric* 53, no. 3 (2020): 241, https://doi.org/10.5325/philrhet.53.3.0239.

15. Philip C. Wander, "The Rhetoric of Science," *Western Speech Communication* 40, no. 4 (1976): 227, https://doi.org/10.1080/10570317609373907.

16. Emily Sohn, "Secrets to Writing a Winning Grant," *Nature* 577, no. 7788 (2019): 133–35, https://doi.org/10.1038/d41586-019-03914-5.

17. For example, in an article on brain sex research, Jordynn Jack noted that by defining null hypotheses as "no sex differences," scientists were implicitly promoting cognitive differences between men and women and thereby influenced the creation of results that affirmed these differences. Jordynn Jack, "How Good Brain Science Gets That Way: Reclaiming the Scientific Study of Sexed and Gendered Brains," in *Feminist Rhetorical Science Studies,* ed. Amanda K. Booher and Julie Jung (Carbondale: Southern Illinois University Press, 2018), 164–82; see also Kenneth Burke, *Language as Symbolic Action: Essays on Life, Literature, and Method* (Berkeley: University of California Press, 1966), 45.

18. Many narrative scholars draw theoretical distinctions between the terms *narrative, story,* and *storytelling.* In the context of science and the environment, Dylan Harris, for example, argues that narratives are "systems of stories" where individual stories inform narratives as "overarching messages." Mithra Moezzi and colleagues view a story as an "object," whereas storytelling is the "performance" of the story for an audience. For the sake of simplicity, I use *story* and *narrative* interchangeably to refer to the circulating discourses about scientific topics and *storytelling* to refer to the practice of science communication. Dylan M. Harris, "Telling the Story of Climate Change: Geologic Imagination, Praxis, and Policy," *Energy Research & Social Science* 31 (2017): 182, https://doi.org/10.1016/j.erss.2017.05.027; Mithra Moezzi, Kathryn B. Janda, and Sea Rotmann, "Using Stories, Narratives, and Storytelling in Energy and Climate Change Research," *Energy Research & Social Science* 31 (2017): 3, https://doi.org/10.1016/j.erss.2017.06.034.

19. Martin Kreiswirth, "Merely Telling Stories? Narrative and Knowledge in the Human Sciences," *Poetics Today* 21, no. 2 (2000): 294, https://doi.org/10.1215/03335372-21-2-293.

20. Kreiswirth, 294.

21. Chris Ingraham, "The Scope and Autonomy of Personal Narrative," *Written Communication* 34, no. 1 (2017): 61, https://doi.org/10.1177/0741088316683147.

22. Kreiswirth, "Merely Telling Stories?," 294.

23. H. Porter Abbott, *The Cambridge Introduction to Narrative* (Cambridge: Cambridge University Press, 2008), 1.

24. Walter R. Fisher, "Narration as a Human Communication Paradigm: The Case of Public Moral Argument," *Communication Monographs* 51, no. 1 (1984): 2.

25. Thomas Basbøll, "A Scientific Paper Shouldn't Tell a Good Story but Present a Strong Argument," *Impact of Social Sciences* (blog), June 1, 2018, https://blogs.lse.ac.uk/impactofsocialsciences/2018/06/01/a-scientific-paper-shouldnt-tell-a-good-story-but-present-a-strong-argument/.

26. Olav Muurlink and Peter McAllister, "Narrative Risks in Science Writing for the Lay Public," *Journal of Science Communication* 14, no. 3 (2015): 1–17, https://doi.org/10.22323/2.14030201; Michael F. Dahlstrom and Shirley S. Ho, "Ethical Considerations of Using Narrative to Communicate Science," *Science Communication* 34, no. 5 (2012): 592–617, https://doi.org/10.1177/1075547012454597.

27. Fisher, "Narration as a Human Communication Paradigm," 1.

28. Arthur C. Graesser, Brent Olde, and Bianca Klettke, "How Does the Mind Construct and Represent Stories?," in *Narrative Impact: Social and Cognitive Foundations,* ed. Melanie Green, Jeffrey Strange, and Timothy Brock (Mahwah, NJ: Erlbaum, 2002), 229–62; Sheila Murphy, Stephanie Demetriades, and Nathan Walter, "Just a Spoonful of Sugar Helps the Messages Go Down: Using Stories and Vicarious Self-Affirmation to Reduce e-Cigarette Use," *Health Communication* 34, no. 3 (2019): 352–60, https://doi.org/10.1080/10410236.2017.1407275; Rick Busselle and Helena Bilandzic, "Fictionality and Perceived Realism in Experiencing Stories: A Model of Narrative Comprehension and Engagement," *Communication Theory* 18, no. 2 (2008): 255–80, https://doi.org/10.1111/j.1468-2885.2008.00322.x; Marcel Machill, Sebastian Köhler, and Markus Waldhauser, "The Use of Narrative Structures in Television News: An Experiment in Innovative Forms of Journalistic Presentation," *European Journal of Communication* 22, no. 2 (2007): 185–205, https://doi.org/10.1177/0267323107076769.

29. Michael F. Dahlstrom, "The Role of Causality in Information Acceptance in Narratives: An Example from Science Communication," *Communication Research* 37, no. 6 (2010): 857–75, https://doi.org/10.1177/0093650210362683; Angeline Sangalang and Emma Frances Bloomfield, "Mother Goose and Mother Nature: Designing Stories to Communicate Information about Climate Change," *Communication Studies* 69, no. 5 (2018): 583–604, https://doi.org/10.1080/10510974.2018.1489872; Sheila T. Murphy et al., "Narrative versus Nonnarrative: The Role of Identification, Transportation, and Emotion in Reducing Health Disparities," *Journal of Communication* 63, no. 1 (2013): 116–37, https://doi.org/10.1111/jcom.12007; Hyun Suk Kim et al., "Narrative Health Communication and Behavior Change: The Influence of Exemplars in the News on Intention to Quit Smoking," *Journal of Communication* 62, no. 3 (2012): 473–92, https://doi.org/10.1111/j.1460-2466.2012.01644.x; Michael D. Slater and Donna Rouner, "Entertainment-Education and Elaboration Likelihood: Understanding the Processing of Narrative Persuasion," *Communication Theory* 12, no. 2 (2002): 173–91, https://doi.org/10.1111/j.1468-2885.2002.tb00265.x; Emily Moyer-Gusé and Robin L. Nabi, "Explaining the Effects of Narrative in an Entertainment Television Program: Overcoming Resistance to Persuasion," *Human Communication Research* 36, no. 1 (2010): 26–52, https://doi.org/10.1111/j.1468-2958.2009.01367.x.

30. Matthew Z. Dudley et al., "The Use of Narrative in Science and Health Communication: A Scoping Review," *Patient Education and Counseling* 112 (2023): 9, https://doi.org/10.1016/j.pec.2023.107752.

31. Dudley et al., 9.

32. Murphy et al., "Narrative versus Nonnarrative," 116–37; Sangalang and Bloomfield, "Mother Goose and Mother Nature," 583–604; Nathan Walter, Sheila T. Murphy, and Traci K. Gillig, "To Walk a Mile in Someone Else's Shoes: How Narratives Can Change Causal Attribution through Story Exploration and Character Customization," *Human Communication Research* 44, no. 1 (2018): 31–57, https://doi.org/10.1093/hcre.12112.

33. P. Sol Hart and Erik C. Nisbet, "Boomerang Effects in Science Communication: How Motivated Reasoning and Identity Cues Amplify Opinion Polarization about Climate Mitigation Policies," *Communication Research* 39, no. 6 (2012): 701–23, https://doi.org/10.1177/0093650211416646; Jack Zhou, "Boomerangs versus Javelins: How Polarization Constrains Communication on Climate Change," *Environmental Politics* 25, no. 5 (2016): 788–811, https://doi.org/10.1080/09644016.2016.1166602.

34. Sangalang and Bloomfield, "Mother Goose and Mother Nature," 583–604.

35. Kim et al., "Narrative Health Communication and Behavior Change," 473–92.

36. Kim Krisberg, "New Movie Puts Public Health, Infectious Disease in Spotlight: Behind the Scenes of 'Contagion,'" *Nation's Health* 41, no. 7 (2011): 1–10, https://www.thenationshealth.org/content/41/7/1.4.

37. Fisher, "Narration as a Human Communication Paradigm," 1–22; Chaïm Perelman, "How Do We Apply Reason to Values?," *Journal of Philosophy* 52, no. 26 (1955): 797–802, https://doi.org/10.2307/2022489.

38. Abbott, *Cambridge Introduction to Narrative*, 40.

39. Fisher, "Narration as a Human Communication Paradigm," 15.

40. Hertha D. Wong, "Pre-Literate Native American Autobiography: Forms of Personal Narrative," *MELUS* 14, no. 1 (1987): 17–32, https://doi.org/10.2307/467470.

41. Ku'Ualoha Ho'Omanawanui, "Hä, Mana, Leo (Breath, Spirit, Voice): Kanaka Maoli Empowerment through Literature," *American Indian Quarterly* 28, no. 1/2 (2004): 86–91, https://www.jstor.org/stable/4139042.

42. Nicole Barile, "Hollywood, Bollywood, Nollywood: Film Industries around the World," *Future of Global Work* (blog), February 26, 2019, https://futureofglobalwork.com/blog/2019/2/26/hollywood-bollywood-nollywood-film-industries-around-the-world.

43. John A. Lynch, *The Origins of Bioethics: Remembering When Medicine Went Wrong* (East Lansing: Michigan State University Press, 2019); Chanda Prescod-Weinstein, "Making Black Women Scientists under White Empiricism:

The Racialization of Epistemology in Physics," *Signs: Journal of Women in Culture and Society* 45, no. 2 (2020): 421–47, https://doi.org/10.1086/704991; Yascha Mounk, "What an Audacious Hoax Reveals about Academia," *The Atlantic*, October 5, 2018, https://www.theatlantic.com/ideas/archive/2018/10/new-sokal-hoax/572212/.

44. John Lyne and Henry F. Howe, "'Punctuated Equilibria': Rhetorical Dynamics of a Scientific Controversy," *Quarterly Journal of Speech* 72, no. 2 (1986): 132–47, https://doi.org/10.1080/00335638609383764.

45. Lauren R. Kolodziejski, "Harms of Hedging in Scientific Discourse: Andrew Wakefield and the Origins of the Autism Vaccine Controversy," *Technical Communication Quarterly* 23, no. 3 (2014): 165–83, https://doi.org/10.1080/10572252.2013.816487.

46. By way of comparison, I draw on the writings of St. Augustine, who sought to leverage the art of rhetoric for evangelization. Augustine argued that Christians had prematurely abandoned rhetoric because of its use by pagans, atheists, and other perceived enemies of the church. Instead, Augustine proposed that rhetoric was a tool, neutral in and of itself but infused with the values and intentions of its wielder. To reject the use of rhetoric wholeheartedly in all circumstances, therefore, was to sacrifice the ability of those who would wield rhetoric for good. Augustine, "On Christian Doctrine," in *The Rhetorical Tradition: Readings from Classical Times to the Present*, ed. Bruce Bizzell and Patricia Herzberg (Boston: Bedford/St. Martin's, 2001), 456.

47. Stephen D. O'Leary, *Arguing the Apocalypse: A Theory of Millennial Rhetoric* (New York: Oxford University Press, 1998), 44.

48. Thomas F. Gieryn, "Boundary-Work and the Demarcation of Science from Non-Science: Strains and Interests in Professional Ideologies of Scientists," *American Sociological Review* 48, no. 6 (1983): 781–95, https://doi.org/10.2307/2095325.

49. Mohammed A. Mostajo-Radji, "Pseudoscience in the Times of Crisis: How and Why Chlorine Dioxide Consumption Became Popular in Latin America during the COVID-19 Pandemic," *Frontiers in Political Science* 3 (2021), https://www.frontiersin.org/articles/10.3389/fpos.2021.621370; Emma Frances Bloomfield and Stephanie Willes, "Religious Masking and the Rhetorical Strategies of Digital Anti-Vaccination Churches," *Western Journal of Communication* (2023), https://doi.org/10.1080/10570314.2023.2174384; Emiliano Rodríguez Mega, "Latin America's Embrace of an Unproven COVID Treatment Is Hindering Drug Trials," *Nature* 586, no. 7830 (2020): 481–83, https://www.nature.com/articles/d41586-020-02958-2.

50. Prescod-Weinstein, "Making Black Women Scientists under White Empiricism," 421.

51. Natasha N. Jones, "My Testimony: Black Feminist Thought in Scientific Communication," in *Routledge Handbook of Scientific Communication*, ed. Cris-

tina Hanganu-Bresch, Michael J. Zerbe, Gabriel Cutrufello, and Stefania M. Maci (New York: Routledge, 2022), 58–68.

52. Christopher Scott Thomas, "Motherhood and Environmental Justice: Women's Environmental Communication, Maternity, and the Water Crisis in Flint, Michigan," in *The Rhetoric of Social Movements: Networks, Power, and New Media,* ed. Nathan Crick (New York: Routledge, 2020), 284–98.

53. Emma Frances Bloomfield and Sara C. VanderHaagen, "Where Women Scientists Belong: Placing Feminist Memory in Biography Collections for Children," *Women's Studies in Communication* 45, no 2. (2022): 187–209, https://doi.org/10.1080/07491409.2021.1941464.

54. Ethan Siegel, "These 5 Women Deserved, and Were Unjustly Denied, a Nobel Prize in Physics," *Forbes,* October 11, 2018, https://www.forbes.com/sites/startswithabang/2018/10/11/these-5-women-deserved-and-were-unjustly-denied-a-nobel-prize-in-physics/.

55. Heather Lazrus et al., "Culture Change to Address Climate Change: Collaborations with Indigenous and Earth Sciences for More Just, Equitable, and Sustainable Responses to Our Climate Crisis," *PLOS Climate* 1, no. 2 (2022): https://doi.org/10.1371/journal.pclm.0000005.

56. Michelle Montgomery and Paulette Blanchard, "Testing Justice: New Ways to Address Environmental Inequalities," *Solutions Journal* (2021): https://thesolutionsjournal.com/2021/03/01/testing-justice-new-ways-to-address-environmental-inequalities/; Kimmerer, *Braiding Sweetgrass*; Sonya Atalay, "Indigenous Science for a World in Crisis," *Public Archaeology* (2020): 1–16, https://doi.org/10.1080/14655187.2020.1781492.

57. Thomas S. Kuhn, *The Structure of Scientific Revolutions,* 3rd ed. (Chicago: University of Chicago Press, 1996), 4.

58. Kuhn, 4.

59. Kuhn, 4, 6.

60. Allan Franklin and Ronald Laymon, *Measuring Nothing, Repeatedly: Null Experiments in Physics* (San Rafael, CA: Morgan & Claypool, 2019).

61. Emma Frances Bloomfield, *Communication Strategies for Engaging Climate Skeptics: Religion and the Environment* (New York: Routledge, 2019).

62. James Phelan, "Rhetoric, Ethics, and Narrative Communication: Or, from Story and Discourse to Authors, Resources, and Audiences," *Soundings: An Interdisciplinary Journal* 94, no. 1/2 (2011): 56, https://www.jstor.org/stable/41200942.

63. Abbott, *Cambridge Introduction to Narrative,* 25.

64. Constraints might include material limits on the channel or medium the story is being shared through, time limits, or technological capabilities. Constraints might also include social restrictions of the space or audience, an institution someone is representing, or one's position that might influence one's style or performance to meet expectations.

65. John Louis Lucaites and Celeste Michelle Condit, "Re-Constructing Narrative Theory: A Functional Perspective," *Journal of Communication* 35, no. 4 (1985): 92, https://doi.org/10.1111/j.1460-2466.1985.tb02975.x.

66. Lucaites and Condit, 92.

67. Fisher, "Narration as a Human Communication Paradigm," 10.

68. Leigh Bernacchi and Tarla Rai Peterson, "How Reductive Scientific Narratives Constrain Possibilities for Citizen Engagement in Community-Based Conservation," in *Environmental Communication and Community,* ed. Tarla Rai Peterson et al. (New York: Routledge, 2016); Karl Dudman and Sara de Wit, "An IPCC That Listens: Introducing Reciprocity to Climate Change Communication," *Climatic Change* 168, no. 1 (2021): 1–12, https://doi.org/10.1007/s10584-021-03186-x.

69. My article discusses Tarana Burke's TED talk, "Me Too is a movement, not a moment," which outlines systemic problems such as media while also outlining how sexual violence is more likely to happen to trans women, Indigenous women, and people with disabilities. Emma Frances Bloomfield, "Rhetorical Constellations and the Inventional/Intersectional Possibilities of #MeToo," *Journal of Communication Inquiry* 43, no. 4 (2019): 410, https://doi.org/10.1177/0196859919866444.

70. Maria Novotny, "Cultural Rhetorics in Precarious Times," *Writing & Rhetoric MKE,* 16 (2020): 1–3, https://dc.uwm.edu/writingmke/16.

71. Leanne Betasamosake Simpson, *As We Have Always Done: Indigenous Freedom through Radical Resistance* (Minneapolis: University of Minnesota Press, 2017), 212.

72. Simpson, 215.

73. Vox Jo Hsu, *Constellating Home: Trans and Queer Asian American Rhetorics* (Columbus: Ohio State University Press, 2022), 9.

74. Sherri Goodman and Pauline Baudu, "Climate Change as a 'Threat Multiplier': History, Uses and Future of the Concept," *Center for Climate & Security,* January 3, 2023, https://councilonstrategicrisks.org/wp-content/uploads/2023/01/38-CCThreatMultiplier.pdf.

75. Philip Eubanks, *The Troubled Rhetoric and Communication of Climate Change: The Argumentative Situation* (New York: Routledge, 2015); Thomas M. Lessl, *Rhetorical Darwinism* (Waco, TX: Baylor University Press, 2012); Heidi Yoston Lawrence, *Vaccine Rhetorics* (Columbus: Ohio State University Press, 2020); H. Dan O'Hair and Mary John O'Hair, eds., *Communicating Science in Times of Crisis: COVID-19 Pandemic* (Hoboken, NJ: Wiley Blackwell, 2021).

76. Michael Calvin McGee, "Text, Context, and the Fragmentation of Contemporary Culture," *Western Journal of Speech Communication* 54, no. 3 (1990): 274–89, https://doi.org/10.1080/10570319009374343.

77. Fisher, "Narration as a Human Communication Paradigm," 14.

78. Chris Russill, "Tipping Point Forewarnings in Climate Change Communication: Some Implications of an Emerging Trend," *Environmental Communication* 2, no. 2 (2008): 133–53, https://doi.org/10.1080/17524030802141711; Gordon R. Sullivan et al., "National Security and the Threat of Climate Change," *CNA Corporation,* 2007, https://www.cna.org/archive/CNA_Files/pdf/national%20security%20and%20the%20threat%20of%20climate%20change.pdf; Jay Gulledge et al., "The Age of Consequences: The Foreign Policy and National Security Implications of Global Climate Change," *Center for Strategic & International Studies,* November 5, 2007, https://www.csis.org/analysis/age-consequences.

79. Catalina M. de Onís, "Energy Colonialism Powers the Ongoing Unnatural Disaster in Puerto Rico," *Frontiers in Communication* 3 (2018), https://doi.org/10.3389/fcomm.2018.00002; Danielle Endres, "The Rhetoric of Nuclear Colonialism: Rhetorical Exclusion of American Indian Arguments in the Yucca Mountain Nuclear Waste Siting Decision," *Communication and Critical/Cultural Studies* 6, no. 1 (2009): 39–60, https://doi.org/10.1080/14791420802632103.

80. Naomi Oreskes, "The Scientific Consensus on Climate Change: How Do We Know We're Not Wrong?," in *Climate Modelling: Philosophical and Conceptual Issues,* ed. Elisabeth A. Lloyd and Eric Winsberg (Cham, Switzerland: Palgrave Macmillan, 2018), 31–64, https://doi.org/10.1007/978-3-319-65058-6_2; Peter T. Doran and Maggie Kendall Zimmerman, "Examining the Scientific Consensus on Climate Change," *Eos* 90, no. 3 (2009): 22–23, https://doi.org/10.1029/2009EO030002; John Cook et al., "Quantifying the Consensus on Anthropogenic Global Warming in the Scientific Literature," *Environmental Research Letters* 8, no. 2 (2013), https://doi.org/10.1088/1748-9326/8/2/024024.

81. For example, Bloomfield, *Communication Strategies for Engaging Climate Skeptics;* Riley E. Dunlap and Aaron M. McCright, "Climate Change Denial: Sources, Actors and Strategies," in *Routledge Handbook of Climate Change and Society,* ed. Constance Lever-Tracy (New York: Routledge, 2010): 240–59; Jean-Daniel Collomb, "The Ideology of Climate Change Denial in the United States," *European Journal of American Studies* 9 (2014): 1–21, https://doi.org/10.4000/ejas.10305.

82. Jason R. Wiles, "Overwhelming Scientific Confidence in Evolution and Its Centrality in Science Education—and the Public Disconnect," *Science Education Review* 9, no. 1 (2010): 18–27, https://files.eric.ed.gov/fulltext/EJ896065.pdf; Randy Moore, "The Revival of Creationism in the United States," *Journal of Biological Education* 35, no. 1 (2000): 17–21, https://doi.org/10.1080/00219266.2000.9655730.

83. Brian McCammack, "Hot Damned America: Evangelicalism and the Climate Change Policy Debate," *American Quarterly* 59, no. 3 (2007): 645–68, https://doi.org/10.1353/aq.2007.0065; Emma Frances Bloomfield, "Rhetorical Strategies in Contemporary Responses to Science and Modernity: Legitimizing

Religion in Human Origins and Climate Change Controversies" (PhD diss., University of Southern California, 2016), http://search.proquest.com/docview/2070186953/?pq-origsite = primo; Cynthia Leifer, "The Role of Doubt in Science," *Pacific Standard,* May 3, 2017, https://psmag.com/news/doubt-has-a-purpose-but-that-doesnt-make-it-ok-to-question-climate-change-evolution-or-vaccines.

84. National Science Teachers Association, "NSTA Position Statement: Evolution," accessed April 14, 2020, https://www.nsta.org/about/positions/evolution.aspx.

85. Cary Funk, Greg Smith, and David Masci, "How Many Creationists Are There in America?," *Scientific American,* February 12, 2019, https://blogs.scientificamerican.com/observations/how-many-creationists-are-there-in-america/.

86. Kolodziejski, "Harms of Hedging in Scientific Discourse," 165.

87. Angus Liu, "In Worrying Trend, Vaccination Rates in U.S. Children Keep Dropping: CDC Reports," *Fierce Pharma,* October 16, 2018, https://www.fiercepharma.com/vaccines/worrying-trend-vaccination-rates-u-s-children-keep-dropping-cdc-reports.

88. Claire Felter, "Measles and the Threat of the Anti-Vaccination Movement," *Council on Foreign Relations,* March 12, 2019, https://www.cfr.org/article/measles-and-threat-anti-vaccination-movement.

89. For example, Felter; IFLScience, "One Map Sums Up the Damage Caused by the Anti-Vaccination Movement," October 24, 2014, https://www.iflscience.com/health-and-medicine/one-map-sums-damage-caused-anti-vaccination-movement/; Azhar Hussain et al., "The Anti-Vaccination Movement: A Regression in Modern Medicine," *Cureus* 10, no. 7 (July 3, 2018), https://doi.org/10.7759/cureus.2919.

90. World Health Organization, "Ten Health Issues WHO Will Tackle This Year," 2019, https://www.who.int/emergencies/ten-threats-to-global-health-in-2019.

91. Data was retrieved from the World Health Organization COVID-19 dashboard on April 12, 2023. World Health Organization, "WHO Coronavirus (COVID-19) Dashboard," April 12, 2023, https://covid19.who.int/; World Health Organization, "Global > United States of America," April 12, 2023, https://covid19.who.int/region/amro/country/us.

92. Jeanne Fahnestock, "Accommodating Science: The Rhetorical Life of Scientific Facts," *Written Communication* 15, no. 3 (1998): 331.

CHAPTER 1

1. Jess Henig, "Climategate," *FactCheck.Org* (blog), December 10, 2009, https://www.factcheck.org/2009/12/climategate/.

2. John Costella, "The Climategate Emails," Lavoisier Group, 2010, https://www.lavoisier.com.au/articles/greenhouse-science/climate-change/climategate-emails.pdf.

3. Henig, "Climategate."

4. John Lyne and Henry F. Howe, "'Punctuated Equilibria': Rhetorical Dynamics of a Scientific Controversy," *Quarterly Journal of Speech* 72, no. 2 (1986): 140, https://doi.org/10.1080/00335638609383764.

5. G. Thomas Goodnight, "Public Discourse," *Critical Studies in Media Communication* 4, no. 4 (1987): 428–32, https://doi.org/10.1080/15295038709360154.

6. David Adam, "Climategate Scientists Cleared of Manipulating Data on Global Warming," *The Guardian,* July 7, 2010, https://www.theguardian.com/environment/2010/jul/08/muir-russell-climategate-climate-science.

7. "What Do the 'Climategate' Hacked CRU Emails Tell Us?," *Skeptical Science,* July 2015, https://www.skepticalscience.com/Climategate-CRU-emails-hacked.htm.

8. E. Calvin Beisner, "Three Great Legacies of Climategate," Cornwall Alliance, December 3, 2014, https://cornwallalliance.org/2014/12/three-great-legacies-of-climategate/; James Delingpole, "Climategate: The Final Nail in the Coffin of 'Anthropogenic Global Warming'?," *Telegraph Blogs,* November 20, 2009, http://blogs.telegraph.co.uk/news/jamesdelingpole/100017393/climategate-the-final-nail-in-the-coffin-of-anthropogenic-global-warming/.

9. Robin McKie, "Climategate 10 Years On: What Lessons Have We Learned?," *The Observer,* November 9, 2019, https://www.theguardian.com/theobserver/2019/nov/09/climategate-10-years-on-what-lessons-have-we-learned.

10. Paul Offit, *Autism's False Prophets: Bad Science, Risky Medicine, and the Search for a Cure* (New York: Columbia University Press, 2010).

11. Lauren R. Kolodziejski, "Harms of Hedging in Scientific Discourse: Andrew Wakefield and the Origins of the Autism Vaccine Controversy," *Technical Communication Quarterly* 23, no. 3 (2014): 165–83, https://doi.org/10.1080/10572252.2013.816487.

12. Walter R. Fisher, "Narration as Human Communication Paradigm: The Case of Public Moral Argument," *Communication Monographs* 51, no. 1 (1984): 14, https://doi.org/10.1080/03637758409390180.

13. The term *disingenuous* is not meant to imply that all rival storytellers are insincere or are not genuine. Instead, the term draws from Kelly and Hoerl's work on Answers in Genesis's Creation Museum, which refers to the space as promoting a "disingenuous scientific controversy." They define a disingenuous scientific controversy as one that "undermine[s] existing scientific knowledge and legitimize[s] pseudoscientific beliefs" (126). I thus use the term to signal rival stories that undermine institutionalized scientific knowledge in ways that are harmful and offer alternative beliefs that may not be considered scientific. Casey Ryan Kelly and Kristen E. Hoerl, "Genesis in Hyperreality: Legitimizing

Disingenuous Controversy at the Creation Museum," *Argumentation and Advocacy* 48, no. 3 (2012): 123–41, https://doi.org/10.1080/00028533.2012.11821759.

14. Naomi Oreskes and Erik M. Conway, *Merchants of Doubt: How a Handful of Scientists Obscured the Truth on Issues from Tobacco Smoke to Global Warming* (New York: Bloomsbury Press, 2011), 34.

15. Steven Schwarze, "Environmental Melodrama," *Quarterly Journal of Speech* 92, no. 3 (2006): 239–61. https://doi.org/10.1080/00335630600938609.

16. Aja Y. Martinez, "A Plea for Critical Race Theory Counterstory: Stock Story versus Counterstory Dialogues Concerning Alejandra's 'Fit' in the Academy," *Composition Studies* 42, no. 2 (2014): 33–55, https://doi.org/10.37514/ATD-B.2016.0933.2.03.

17. Martinez, 38.

18. Martinez, 38.

19. Martinez, 38.

20. Martinez, 51.

21. Xiaoquan Zhao et al., "Attention to Science/Environment News Positively Predicts and Attention to Political News Negatively Predicts Global Warming Risk Perceptions and Policy Support," *Journal of Communication* 61, no. 4 (2011): 713, https://doi.org/10.1111/j.1460-2466.2011.01563.x.

22. Intergovernmental Panel on Climate Change, Working Group II, "Climate Change 2022: Impacts, Adaptation and Vulnerability," February 28, 2022, https://www.ipcc.ch/report/sixth-assessment-report-working-group-ii/.

23. This consensus has been well established by a series of articles based on abstracts of peer-reviewed scientific reports and interviews with experts in environmental science and climatology. Of additional note is a study by Bray, who offers that those who dissent from Intergovernmental Panel on Climate Change reports may do so on the grounds that they disagree with the underestimation of IPCC data and not the presence and severity of climate change. William R. L. Anderegg et al., "Expert Credibility in Climate Change," *Proceedings of the National Academy of Sciences* (2010), https://doi.org/10.1073/pnas.1003187107; Dennis Bray, "The Scientific Consensus of Climate Change Revisited," *Environmental Science & Policy* 13, no. 5 (2010): 340–50, https://doi.org/10.1016/j.envsci.2010.04.001; John Cook et al., "Consensus on Consensus: A Synthesis of Consensus Estimates on Human-Caused Global Warming," *Environmental Research Letters* 11, no. 4 (2016), https://doi.org/10.1088/1748-9326/11/4/048002; Peter T. Doran and Maggie Kendall Zimmerman, "Examining the Scientific Consensus on Climate Change," *Eos* 90, no. 3 (2009): 22–23, https://doi.org/10.1029/2009EO030002; Naomi Oreskes, "The Scientific Consensus on Climate Change: How Do We Know We're Not Wrong?," in *Climate Modelling: Philosophical and Conceptual Issues*, ed. Elisabeth A. Lloyd and Eric Winsberg (Cham, Switzerland: Palgrave Macmillan, 2018), 31–64.

24. Anthony Leiserowitz et al., "Global Warming's Six Americas," Yale Program on Climate Change Communication, September 2021, https://climatecommunication.yale.edu/about/projects/global-warmings-six-americas/.

25. Anthony Leiserowitz et al., "Climate Change in the American Mind: April 2020," Yale Program on Climate Change Communication, April 2020, https://climatecommunication.yale.edu/publications/climate-change-in-the-american-mind-april-2020/.

26. Leiserowitz et al., "Climate Change in the American Mind."

27. Leah Ceccarelli, "Manufactured Scientific Controversy: Science, Rhetoric, and Public Debate," *Rhetoric & Public Affairs* 14, no. 2 (2011): 195–228, https://doi.org/10.1353/rap.2010.0222; Oreskes and Conway, *Merchants of Doubt.*

28. Emma Frances Bloomfield, *Communication Strategies for Engaging Climate Skeptics: Religion and the Environment* (New York: Routledge, 2019); David Domke and Kevin Coe, *The God Strategy: How Religion Became a Political Weapon in America* (Oxford: Oxford University Press, 2010); Robin Globus Veldman, *The Gospel of Climate Skepticism* (Oakland: University of California Press, 2019).

29. For example, the Pew Research Center reported that 90 percent of Democrats polled believe that "the federal government is doing too little to reduce the effects of climate change," while only 39 percent of Republicans agreed with this same statement (see Funk and Hefferon). In my first book, I interviewed some Republican environmentalists who used the language of "coming out" to describe what they felt were unfair assumptions about how Republicans viewed the environment. Cary Funk and Meg Hefferon, "U.S. Public Views on Climate and Energy," *Pew Research Center Science & Society* (blog), November 25, 2019, https://www.pewresearch.org/science/2019/11/25/u-s-public-views-on-climate-and-energy/; Bloomfield, *Communication Strategies for Engaging Climate Skeptics.*

30. Derek G. Ross, "Common Topics and Commonplaces of Environmental Rhetoric," *Written Communication* 30, no. 1 (2013): 91–131, https://doi.org/10.1177/0741088312465376.

31. It is important to note that the relationship between Christianity and climate skepticism is not deterministic. Some Christians use their faith as justification and motivation for environmental protection. For more information on green faith or "creation care" initiatives, see Emma Frances Bloomfield, "The Reworking of Evangelical Christian Ecocultural Identity in the Creation Care Movement," in *The Routledge Handbook of Ecocultural Identity,* ed. Tema Milstein and José Castro-Sotomayor (New York: Routledge, 2020), 195–207; Katharine K. Wilkinson, *Between God & Green: How Evangelicals Are Cultivating a Middle Ground on Climate Change* (Oxford: Oxford University Press, 2012); Stephen Ellingson, *To Care for Creation: The Emergence of the Religious Environmental Movement* (Chicago: University of Chicago Press, 2016).

32. Cornwall Alliance, "The Biblical Perspective on Environmental Stewardship: Subduing and Ruling the Earth to the Glory of God and the Benefit of Our Neighbors," accessed April 2, 2018, https://cornwallalliance.org/landmark-documents/the-biblical-perspective-of-environmental-stewardship-subduing-and-ruling-the-earth-to-the-glory-of-god-and-the-benefit-of-our-neighbors/.

33. David C. Barker and David H. Bearce, "End-Times Theology, the Shadow of the Future, and Public Resistance to Addressing Global Climate Change," *Political Research Quarterly* 66, no. 2 (2013): 267–79, https://doi.org/10.1177/1065912912442243.

34. Robert Asen, "Neoliberalism, the Public Sphere, and a Public Good," *Quarterly Journal of Speech* 103, no. 4 (2017): 329–49, https://doi.org/10.1080/00335630.2017.1360507; Emma Frances Bloomfield, "The Rhetoric of Energy Darwinism: Neoliberal Piety and Market Autonomy in Economic Discourse," *Rhetoric Society Quarterly* 49, no. 4 (2019): 320–41, https://doi.org/10.1080/02773945.2019.1634831.

35. Bloomfield, "The Rhetoric of Energy Darwinism," 320–41.

36. Tema Milstein, "Nature Identification: The Power of Pointing and Naming," *Environmental Communication* 5, no. 1 (2011): 3–24, https://doi.org/10.1080/17524032.2010.535836.

37. Kevin Michael DeLuca, "Unruly Arguments: The Body Rhetoric of Earth First!, ACT UP, and Queer Nation," *Argumentation and Advocacy* 36, no. 1 (1999): 12–13, https://doi.org/10.1080/00028533.1999.11951634.

38. Jessica Hernandez, *Fresh Banana Leaves: Healing Indigenous Landscapes through Indigenous Science* (Berkeley, CA: North Atlantic Books, 2022), 32.

39. Milstein, "Nature Identification," 3–24; Donovan S. Conley and Lawrence J. Mullen, "Righting the Commons in Red Rock Canyon," *Communication and Critical/Cultural Studies* 5, no. 2 (2008): 180–99, https://doi.org/10.1080/14791420801989694; Emma Frances Bloomfield, "Transcorporeal Identification and Strategic Essentialism in Eco-Horror: *mother!*'s Ecofeminist Rhetorical Strategies," *Environmental Communication* 15, no. 3 (2021): 339–52, https://doi.org/10.1080/17524032.2020.1833059.

40. Ryan Holifield, Jayajit Chakraborty, and Gordon Walker, *The Routledge Handbook of Environmental Justice* (New York: Routledge, 2017).

41. Steven Schwarze et al., "Environmental Melodrama, Coal, and the Politics of Sustainable Energy in *The Last Mountain*," *International Journal of Sustainable Development* 17, no. 2 (2014): 111, https://doi.org/10.1504/IJSD.2014.061776.

42. Daniel R. Wildcat, *Red Alert!: Saving the Planet with Indigenous Knowledge* (Wheat Ridge, CO: Fulcrum, 2009).

43. Hernandez, *Fresh Banana Leaves,* 11.

44. Wanjohi Kabukuru, "Young African Climate Activists Speak Out Ahead of COP27 in Egypt," *PBS,* November 6, 2022, https://www.pbs.org/newshour/world/young-african-climate-activists-speak-out-ahead-of-cop27-in-egypt.

45. Kenya Evelyn, "'Like I Wasn't There': Climate Activist Vanessa Nakate on Being Erased from a Movement," *The Guardian*, January 29, 2020, https://www.theguardian.com/world/2020/jan/29/vanessa-nakate-interview-climate-activism-cropped-photo-davos.

46. Kermit Pattison, "Lucy and Ardi: The Two Fossils That Changed Human History," *BBC Science Focus Magazine*, March 7, 2021, https://www.sciencefocus.com/the-human-body/lucy-and-ardi-the-two-fossils-that-changed-human-history/.

47. The technical definition of *theory* indicates an explanation that has been repeatedly supported through data and experimentation. This is starkly different from the more common, everyday usage of "theory" as a guess, hypothesis, or conjecture that is typically not backed up by evidence. Tia Ghose, "'Just a Theory': 7 Misused Science Words," *Scientific American*, April 2, 2013, https://www.scientificamerican.com/article/just-a-theory-7-misused-science-words/.

48. Pew Research Center, "Overview: The Conflict between Religion and Evolution," February 3, 2014, https://www.pewforum.org/2009/02/04/overview-the-conflict-between-religion-and-evolution/; see also David Masci, "The History of the Evolution Debate in the United States," Pew Research Center, February 6, 2019, https://www.pewforum.org/essay/darwin-in-america/.

49. Michael B. Berkman and Eric Plutzer, "Enablers of Doubt: How Future Teachers Learn to Negotiate the Evolution Wars in Their Classrooms," *ANNALS of the American Academy of Political and Social Science* 658, no. 1 (2015): 255, https://doi.org/10.1177/0002716214557783.

50. John Angus Campbell, "Darwin and the Origin of Species: The Rhetorical Ancestry of an Idea," *Communication Monographs* 37, no. 1 (1970): 1, https://doi.org/10.1080/03637757009375644.

51. Campbell, 9.

52. Randy Moore, "The Revival of Creationism in the United States," *Journal of Biological Education* 35, no. 1 (2000): 17–21, https://doi.org/10.1080/00219266.2000.9655730.

53. Thomas M. Lessl, "Punctuation in the Constitution of Public Identities: Primary and Secondary Sequences in the Scopes Trial," *Communication Theory* 3, no. 2 (1993): 97, https://doi.org/10.1111/j.1468-2885.1993.tb00060.x.

54. Epperson v. Arkansas, 393 U.S. 97 (October 16, 1968): 97; McLean v. State of Arkansas, 663 F2d 47 (October 8, 1981): 47; Edwards v. Aguillard, 482 U.S. 578 (December 10, 1986): 578.

55. Webster v. New Lenox School District No. M, 917 F2d 1004 (February 27, 1990): 1004; Philip Sparr, "Special 'Effects': Kitzmiller v. Dover Area School District, 400 F. Supp. 2d 707 (M.D. Pa. 2005), and the Fate of Intelligent Design in Our Public Schools," *Nebraska Law Review* 86 (2007): 708.

56. Kelly and Hoerl, "Genesis in Hyperreality," 124.

57. Kelly and Hoerl, "Genesis in Hyperreality," 123–41; Emma Frances Bloomfield, "Ark Encounter as Material Apocalyptic Rhetoric: Contemporary Creationist Strategies on Board Noah's Ark," *Southern Communication Journal* 82, no. 5 (2017): 263–77, https://doi.org/10.1080/1041794X.2017.1360384; John Lynch, "'Prepare to Believe': The Creation Museum as Embodied Conversion Narrative," *Rhetoric and Public Affairs* 16, no. 1 (2013): 1–28, https://doi.org/10.14321/rhetpublaffa.16.1.0001; Emma Frances Bloomfield, "Sensory Engagement with the Rhetoric of Science: Creationist *Copia* at the Discovery Center for Science and Earth History," *Rhetoric and Public Affairs* 25, no. 4 (2022): 65–93, https://muse.jhu.edu/article/885165.

58. Creation Network, "Everything You'll Need on Creation in One Place," 2018, https://creationnetwork.org/.

59. Pew Research Center, "Muslim Views on Religion, Science and Popular Culture," April 30, 2013, https://www.pewforum.org/2013/04/30/the-worlds-muslims-religion-politics-society-science-and-popular-culture/; Pew Research Center, "Science and Religion in Central and Eastern Europe," May 10, 2017, https://www.pewforum.org/2017/05/10/science-and-religion/; Pew Research Center, "Latin American Views on Religion and Science," November 13, 2014, https://www.pewforum.org/2014/11/13/chapter-8-religion-and-science/.

60. Jon D. Miller, Eugenie C. Scott, and Shinji Okamoto, "Public Acceptance of Evolution," *Science* 313, no. 5788 (2006): 765–66, https://doi.org/10.1126/science.1126746.

61. Cary Funk et al., "Biotechnology Research Viewed with Caution Globally, but Most Support Gene Editing for Babies to Treat Disease," *Pew Research Center Science & Society* (blog), December 10, 2020, https://www.pewresearch.org/science/2020/12/10/biotechnology-research-viewed-with-caution-globally-but-most-support-gene-editing-for-babies-to-treat-disease/.

62. Funk et al.

63. Pew Research Center, "Major Gaps between the Public, Scientists on Key Issues," July 1, 2015, http://www.pewinternet.org/interactives/public-scientists-opinion-gap/.

64. National Science Board, "Science and Technology: Public Attitudes and Understanding," Science & Engineering Indicators 2018, https://www.nsf.gov/statistics/2018/nsb20181/report/sections/science-and-technology-public-attitudes-and-understanding/public-knowledge-about-s-t.

65. Emma Frances Bloomfield, "Rhetorical Strategies in Contemporary Responses to Science and Modernity: Legitimizing Religion in Human Origins and Climate Change Controversies" (PhD diss., University of Southern California, 2016), http://search.proquest.com/docview/2070186953/?pq-origsite=primo.

66. John Tagliabue, "Pope Bolsters Church's Support for Scientific View of Evolution," *New York Times,* October 25, 1996, https://www.nytimes

.com/1996/10/25/world/pope-bolsters-church-s-support-for-scientific-view-of-evolution.html.

67. Hernandez, *Fresh Banana Leaves,* 32.

68. Zahraa Saiyed and Paul D. Irwin, "Native American Storytelling toward Symbiosis and Sustainable Design," *Energy Research & Social Science* 31 (2017): 250, https://doi.org/10.1016/j.erss.2017.05.029.

69. Norbert Francis, "The Study of Traditional Narrative and the Science of Human Evolution: Complementary Perspectives," *Interdisciplinary Literary Studies* 13, no. 1–2 (2011): 108, https://www.jstor.org/stable/41328515.

70. Although not called *vaccines,* inoculation practices were commonplace before Edward Jenner's innovations in 1796, including the inoculation techniques of Lady Mary Wortley Montagu and Charles Maitland, who brought these practices to England from Turkey, where they were more widely practiced. Diana Barnes, "The Public Life of a Woman of Wit and Quality: Lady Mary Wortley Montagu and the Vogue for Smallpox Inoculation," *Feminist Studies* 38, no. 2 (2012): 330–62, https://www.jstor.org/stable/23269190.

71. Robert M. Wolfe and Lisa K. Sharp, "Anti-Vaccinationists Past and Present," *BMJ* 325, no. 7361 (2002): 430, https://doi.org/10.1136/bmj.325.7361.430.

72. Stephanie Willes, "The Affordances of the Internet and the Creation of Villains and Heroes in Anti-Vaccine Stories: A Narrative Criticism of VaxTruth" (master's thesis, University of Nevada, Las Vegas, 2019).

73. Anna Kata, "Anti-Vaccine Activists, Web 2.0, and the Postmodern Paradigm—An Overview of Tactics and Tropes Used Online by the Anti-Vaccination Movement," *Vaccine* 30, no. 25 (2012): 3778–89, https://doi.org/10.1016/j.vaccine.2011.11.112.

74. Heidi Yoston Lawrence, *Vaccine Rhetorics* (Columbus: Ohio State University Press, 2020); Jennifer A. Reich, *Calling the Shots: Why Parents Reject Vaccines* (New York: NYU Press, 2018).

75. R. J. Reinhart, "Fewer in U.S. Continue to See Vaccines as Important," *Gallup,* January 14, 2020, https://news.gallup.com/poll/276929/fewer-continue-vaccines-important.aspx.

76. Centers for Disease Control, "Measles: Cases and Outbreaks," June 28, 2019, https://www.cdc.gov/measles/cases-outbreaks.html; Centers for Disease Control, "Pertussis and Whooping Cough," February 14, 2019, https://www.cdc.gov/pertussis/countries/index.html.

77. World Health Organization, "Ten Health Issues WHO Will Tackle This Year," 2019, https://www.who.int/emergencies/ten-threats-to-global-health-in-2019.

78. Beth Mostafavi, "The Harm of Vaccine Refusals," *Michigan Health Lab,* April29,2016,https://labblog.uofmhealth.org/body-work/new-study-emphasizes-harm-of-vaccine-refusals.

79. Alice Park, "Doctor behind Vaccine-Autism Link Loses License," *Time,* May 24, 2010, http://healthland.time.com/2010/05/24/doctor-behind-vaccine-autism-link-loses-license/. See Kolodziejski for an extensive breakdown of the errors and conflicts on interest in Wakefield's article: Kolodziejski, "Harms of Hedging in Scientific Discourse," 165–83.

80. Anna Kata, "A Postmodern Pandora's Box: Anti-Vaccination Misinformation on the Internet," *Vaccine* 28, no. 7 (2010): 1709–16, https://doi.org/10.1016/j.vaccine.2009.12.022.

81. Mark A. Largent, *Vaccine: The Debate in Modern America* (Baltimore: Johns Hopkins University Press, 2012), 1.

82. There are many excellent studies examining the digital circulation of antivaccination narratives. In addition to the ones cited directly, interested readers may also seek out the journal *Vaccine*'s special issue on "The Role of Internet Use in Vaccination Decisions." Terry Connolly and Jochen Reb, "Toward Interactive, Internet-Based Decision Aid for Vaccination Decisions: Better Information Alone Is Not Enough," *Vaccine* 30, no. 25 (2012): 3813–18, https://doi.org/10.1016/j.vaccine.2011.12.094; Ashley Shelby and Karen Ernst, "Story and Science," *Human Vaccines & Immunotherapeutics* 9, no. 8 (2013): 1795–1801, https://doi.org/10.4161/hv.24828; Holly O. Witteman and Brian J. Zikmund-Fisher, "The Defining Characteristics of Web 2.0 and Their Potential Influence in the Online Vaccination Debate," *Vaccine* 30, no. 25 (2012): 3734–40, https://doi.org/10.1016/j.vaccine.2011.12.039.

83. Emma Frances Bloomfield and Stephanie Willes, "Religious Masking and the Rhetorical Strategies of Digital Anti-Vaccination Churches," *Western Journal of Communication* (2023), https://doi.org/10.1080/10570314.2023.2174384.

84. James Colgrove, *State of Immunity: The Politics of Vaccination in Twentieth-Century America* (Berkeley: University of California Press, 2006); Bernice L. Hausman et al., "'Poisonous, Filthy, Loathsome, Damnable Stuff': The Rhetorical Ecology of Vaccination Concern," *Yale Journal of Biology and Medicine* 87, no. 4 (2014): 403–16; Gregory A. Poland and Robert M. Jacobson, "The Age-Old Struggle against the Antivaccinationists," *New England Journal of Medicine* 364, no. 2 (2011): 97–99, https://doi.org/10.1056/NEJMp1010594.

85. Heidi Y. Lawrence, "Fear of the Irreparable: Narratives in Vaccination Rhetoric," *Narrative Inquiry in Bioethics* 6, no. 3 (2016): 205–9, https://doi.org/10.1353/nib.2016.0060.

86. Lawrence, "Fear of the Irreparable," 205.

87. Lawrence, "Fear of the Irreparable," 205–9. See also Largent, *Vaccine.*

88. Jordynn Jack, *Autism and Gender: From Refrigerator Mothers to Computer Geeks* (Urbana: University of Illinois Press, 2014); Willes, "The Affordances of the Internet."

89. John A. Lynch, *The Origins of Bioethics: Remembering When Medicine Went Wrong* (East Lansing: Michigan State University Press, 2019).

90. All of these cases are important for the history of science and as context for science communicators. I thus encourage readers to explore these instances of scientific injustice in more detail through the following texts, among others written on these topics: Lynch, *The Origins of Bioethics*; Editorial, "Henrietta Lacks: Science Must Right a Historical Wrong," *Nature* 585, no. 7823 (2020): 7, https://doi.org/10.1038/d41586-020-02494-z; Alexandra Minna Stern, "Forced Sterilization Policies in the US Targeted Minorities and Those with Disabilities—and Lasted into the 21st Century," *The Conversation,* August 26, 2020, http://theconversation.com/forced-sterilization-policies-in-the-us-targeted-minorities-and-those-with-disabilities-and-lasted-into-the-21st-century-143144; Mark A. Largent, *Breeding Contempt: The History of Coerced Sterilization in the United States* (New Brunswick, NJ: Rutgers University Press, 2011).

91. Marco Marani et al., "Intensity and Frequency of Extreme Novel Epidemics," *Proceedings of the National Academy of Sciences* 118, no. 35 (2021), https://doi.org/10.1073/pnas.2105482118.

92. CNN Editorial Research, "Covid-19 Pandemic Timeline Fast Facts," *CNN,* September 4, 2022, https://www.cnn.com/2021/08/09/health/covid-19-pandemic-timeline-fast-facts/index.html.

93. Michael Worobey et al., "The Huanan Seafood Wholesale Market in Wuhan Was the Early Epicenter of the COVID-19 Pandemic," *Science* 377, no. 6609 (2022): 951–59, https://doi.org/10.1126/science.abp8715.

94. CNN Editorial Research, "Covid-19 Pandemic Timeline Fast Facts"; Gianfranco Spiteri et al., "First Cases of Coronavirus Disease 2019 (COVID-19) in the WHO European Region, 24 January to 21 February 2020," *Eurosurveillance* 25, no. 9 (2020): 1–6, https://doi.org/10.2807/1560-7917.ES.2020.25.9.2000178.

95. World Health Organization, "Coronavirus Disease (COVID-19)," May 31, 2021, https://www.who.int/news-room/questions-and-answers/item/coronavirus-disease-covid-19.

96. Kevin Kavanagh, "Is COVID-19 Primarily a Heart and Vascular Disease?," *Infection Control Today,* September 8, 2020, https://www.infectioncontroltoday.com/view/is-covid-19-primarily-a-heart-and-vascular-diseases; World Health Organization, "Coronavirus Disease (COVID-19)."

97. U.S. Department of Defense, "Coronavirus: Timeline," September 14, 2022, https://www.defense.gov/Explore/Spotlight/Coronavirus-DOD-Response/Timeline/.

98. "US Coronavirus Vaccine Tracker," *USAFacts,* April 12, 2023, https://usafacts.org/visualizations/covid-vaccine-tracker-states; "Tracking Coronavirus Vaccinations around the World," *New York Times,* March 13, 2023, https://www.nytimes.com/interactive/2021/world/covid-vaccinations-tracker.html.

99. Mayo Clinic, "Herd Immunity and COVID-19: What You Need to Know," September 27, 2022, https://www.mayoclinic.org/diseases-conditions/coronavirus/in-depth/herd-immunity-and-coronavirus/art-20486808.

100. Hannah Ritchie et al., "Coronavirus Pandemic (COVID-19)," *Our World in Data,* March 5, 2020, https://ourworldindata.org/covid-vaccinations; "Tracking Coronavirus Vaccinations around the World," *New York Times.*

101. This poll was conducted between August 10 and August 16, 2020, during which the United States reported over one hundred and fifty thousand deaths due to COVID-19 and over five million positive cases. Ben Kamisar and Melissa Holzberg, "Poll: Less than Half of Americans Say They'll Get a Coronavirus Vaccine," *NBC News,* August 18, 2020, https://www.nbcnews.com/politics/2020-election/poll-less-half-americans-say-they-ll-get-coronavirus-vaccine-n1236971.

102. Cary Funk and Alec Tyson, "Intent to Get a COVID-19 Vaccine Rises to 60% as Confidence in Research and Development Process Increases," Pew Research Center, December 3, 2020, https://www.pewresearch.org/science/2020/12/03/intent-to-get-a-covid-19-vaccine-rises-to-60-as-confidence-in-research-and-development-process-increases/.

103. Dror Walter, Yotam Ophir, and Hui Ye, "Conspiracies, Misinformation and Resistance to Public Health Measures during COVID-19 in White Nationalist Online Communication," *Vaccine* (2023), https://doi.org/10.1016/j.vaccine.2023.03.050.

104. Diana Daly, "What Are the 'Reopen' Protesters Really Saying?," *The Conversation,* May 1, 2020, http://theconversation.com/what-are-the-reopen-protesters-really-saying-137558.

105. Daly.

106. Karen M. Douglas, "COVID-19 Conspiracy Theories," *Group Processes & Intergroup Relations* 24, no. 2 (2021): 270–75, https://doi.org/10.1177/1368430220982068.

107. Human Rights Watch, "Covid-19 Fueling Anti-Asian Racism and Xenophobia Worldwide," May 12, 2020, https://www.hrw.org/news/2020/05/12/covid-19-fueling-anti-asian-racism-and-xenophobia-worldwide.

108. I purposefully omit specific details about the conspiracy narratives to reduce belief echoes, which refer to how the repetition of conspiracy beliefs can increase their circulation and adoption. Emily Thorson, "Belief Echoes: The Persistent Effects of Corrected Misinformation," *Political Communication* 33, no. 3 (2016): 460–80, https://doi.org/10.1080/10584609.2015.1102187.

109. Todd K. Hartman et al., "Different Conspiracy Theories Have Different Psychological and Social Determinants: Comparison of Three Theories about the Origins of the COVID-19 Virus in a Representative Sample of the UK Population," *Frontiers in Political Science* 3 (2021): 1, https://www.frontiersin.org/articles/10.3389/fpos.2021.642510.

110. Susan M. Reverby, "Racism, Disease, and Vaccine Refusal: People of Color Are Dying for Access to COVID-19 Vaccines," *PLOS Biology* 19, no. 3 (2021): 1–3, https://doi.org/10.1371/journal.pbio.3001167.

111. Centers for Disease Control, "Community, Work, and School," February 11, 2020, https://www.cdc.gov/coronavirus/2019-ncov/community/health-equity/race-ethnicity.html.

112. Cindy Pearson, "We Can't Let Dr. Susan Moore's Death Be in Vain," National Women's Health Network, December 29, 2020, https://nwhn.org/its-up-to-us-not-to-let-dr-susan-moores-death-be-in-vain.

113. Pearson.

114. Thomas S. Kuhn, *The Structure of Scientific Revolutions,* 3rd. ed. (Chicago: University of Chicago Press, 1996).

115. Ceccarelli, "Manufactured Scientific Controversy," 197. For example, Roy Spencer is a meteorologist who is often cited by climate skeptical groups because he questions humanity's influence on the Earth's temperature. Kevin Grandia, "Roy Spencer," *DeSmog* (blog), accessed September 27, 2022, https://www.desmog.com/roy-spencer/.

116. Bloomfield, "Ark Encounter as Material Apocalyptic Rhetoric," 263–77; Kelly and Hoerl, "Genesis in Hyperreality," 123–41; Robert T. Pennock, "Naturalism, Evidence and Creationism: The Case of Phillip Johnson," *Biology and Philosophy* 11, no. 4 (1996): 543–59, https://doi.org/10.1007/BF00138334; Charles Alan Taylor, "Of Audience, Expertise and Authority: The Evolving Creationism Debate," *Quarterly Journal of Speech* 78, no. 3 (1992): 277–95, https://doi.org/10.1080/00335639209383997.

117. Melissa L. Carrion, "'You Need to Do Your Research': Vaccines, Contestable Science, and Maternal Epistemology," *Public Understanding of Science* 27, no. 3 (2018): 310–24, https://doi.org/10.1177/0963662517728024; Will Huntsberry, "Medical Board Charges San Diego Doctor Who's Doled Out Dozens of Vaccine Exemptions," *Voice of San Diego,* October 22, 2019, https://www.voiceofsandiego.org/topics/news/medical-board-charges-san-diego-doctor-whos-doled-out-dozens-of-vaccine-exemptions/; Kolodziejski, "Harms of Hedging in Scientific Discourse," 165–83.

118. Megan Molteni, "The 60-Year-Old Scientific Screwup That Helped Covid Kill," *Wired,* May 13, 2021, https://www.wired.com/story/the-teeny-tiny-scientific-screwup-that-helped-covid-kill/.

119. Oreskes and Conway, *Merchants of Doubt.*

120. Peter J. Jacques, Riley E. Dunlap, and Mark Freeman, "The Organisation of Denial: Conservative Think Tanks and Environmental Scepticism," *Environmental Politics* 17, no. 3 (2008): 349–85, https://doi.org/10.1080/09644010802055576; Riley E. Dunlap and Aaron M. McCright, "Climate Change Denial: Sources, Actors and Strategies," in *Routledge Handbook of Climate Change and Society,* ed. Constance Lever-Tracy (New York: Routledge, 2010), 240–59.

121. Charity Navigator, "Institute for Creation Research," 2013, http://www.charitynavigator.org/index.cfm?bay=search.irs&ein=953523177; Charity

Navigator, "Rating for Answers in Genesis," 2017, http://www.charitynavigator.org/index.cfm?bay=search.summary&orgid=5214; "Ken Ham Can Now Build 'Noah's Ark' Thanks to Bill Nye Creationist Debate," *Huffington Post,* February 28, 2014, http://www.huffingtonpost.com/2014/02/28/ken-ham-noahs-ark_n_4873893.html.

122. Josh Bloom, "Homeopathy Supplement Peddler's Despicable Anti-Vaccine Goldmine," American Council on Science and Health, March 16, 2019, https://www.acsh.org/news/2019/03/16/homeopathy-supplement-peddlers-despicable-anti-vaccine-goldmine-13880; Vincent Iannelli, "Money and Motivation of the Anti-Vaccine Movement," *Vaxopedia,* February 11, 2018, https://vaxopedia.org/2018/02/11/money-and-motivation-of-the-anti-vaccine-movement/; Lena H. Sun and Amy Brittain, "Meet the New York Couple Donating Millions to the Anti-Vax Movement," *Washington Post,* June 19, 2019, https://www.washingtonpost.com/national/health-science/meet-the-new-york-couple-donating-millions-to-the-anti-vax-movement/2019/06/18/9d791bcc-8e28–11e9-b08e-cfd89bd36d4e_story.html.

123. Isaac Stanley-Becker and Tony Romm, "The Anti-Quarantine Protests Seem Spontaneous. But behind the Scenes, a Powerful Network Is Helping," *Washington Post,* April 22, 2020, https://www.washingtonpost.com/politics/inside-the-conservative-networks-backing-anti-quarantine-protests/2020/04/22/da75c81e-83fe-11ea-a3eb-e9fc93160703_story.html; Tina Nguyen, "How a Tea Party–Linked Group Plans to Turbocharge Lockdown Protests," *Politico,* April 24, 2020, https://www.politico.com/news/2020/04/24/tea-party-lockdown-coronavirus-206083.

124. Martinez, "A Plea for Critical Race Theory Counterstory," 52.

CHAPTER 2

1. Kate Taylor, "A Worker Accused of Spoiling More Than 500 Vaccine Doses Is Arrested," *New York Times,* December 31, 2020, https://www.nytimes.com/live/2020/12/31/world/coronavirus-updates.

2. U.S. Department of Justice, "Hospital Pharmacist to Plead Guilty to Attempting to Spoil Hundreds of COVID Vaccine Doses," January 26, 2021, https://www.justice.gov/opa/pr/hospital-pharmacist-plead-guilty-attempting-spoil-hundreds-covid-vaccine-doses.

3. Isaac Stanley-Becker and Andrea Salcedo, "Wisconsin Pharmacist Who 'Intentionally' Spoiled More than 500 Vaccine Doses Is Arrested, Police Say," *Washington Post,* December 31, 2020, https://www.washingtonpost.com/nation/2020/12/31/covid-vaccines-destroyed-wisconsin-hospital/; Bruce Vielmetti, "Judge Declines to Jail Pharmacist in Vaccine Sabotage after Guilty Plea; Sentencing Will Be in June," *Milwaukee Journal Sentinel,* February 9,

2021,https://www.jsonline.com/story/news/crime/2021/02/09/prosecutors-want-pharmacist-jailed-until-sentenced-vaccine-tampering-case/4447741001/.

4. Vielmetti, "Judge Declines to Jail Pharmacist."

5. Vielmetti.

6. Vielmetti.

7. Tommy Beer, "Large Numbers of Health Care and Frontline Workers Are Refusing Covid-19 Vaccine," *Forbes,* January 2, 2021, https://www.forbes.com/sites/tommybeer/2021/01/02/large-numbers-of-health-care-and-frontline-workers-are-refusing-covid-19-vaccine/; Colleen Shalby et al., "Some Healthcare Workers Refuse to Take COVID-19 Vaccine, Even with Priority Access," *Los Angeles Times,* December 31, 2020, https://www.latimes.com/california/story/2020-12-31/healthcare-workers-refuse-covid-19-vaccine-access.

8. Joshua Zitser, "Despite Having Intimate Knowledge of the Pain and Death Caused by the Coronavirus, a Surprising Number of US Healthcare Workers Are Refusing to Get a COVID-19 Vaccine," *Business Insider,* January 10, 2021, https://www.businessinsider.com/covid-19-surprising-number-of-us-healthcare-workers-refuse-vaccines-2021-1.

9. Campbell's structure of mythology outlines twelve stages of a hero's journey venturing out of one's ordinary world into a special world through responding to and conquering an ordeal. Joseph Campbell, *The Hero with a Thousand Faces* (Novato, CA: New World Library, 2008).

10. Jerome S. Bruner, "Life as Narrative," *Social Research* 54, no. 1 (1987): 17, https://www.jstor.org/stable/40970444.

11. This story was used as a reason to reelect Lincoln for his second presidential term by encouraging voters not to change leaders during a tumultuous time in US history.

12. Toolan includes quasi-humans and sentient beings in his definition of narrative under the condition that humans can learn from those characters' experiences. Michael Toolan, *Narrative: A Critical Linguistic Introduction* (New York: Routledge, 2012), 8.

13. Tema Milstein, "Nature Identification: The Power of Pointing and Naming," *Environmental Communication* 5, no. 1 (2011): 3–24, https://doi.org/10.1080/17524032.2010.535836.

14. Jessica Hernandez, *Fresh Banana Leaves: Healing Indigenous Landscapes through Indigenous Science* (Berkeley, CA: North Atlantic Books, 2022); Yukari Kunisue, "Listening to the Natural World: Ecopsychology of Listening from a Hawai'ian Spiritual Perspective," in *Voice and Environmental Communication,* ed. Jennifer Peeples and Stephen Depoe (New York: Palgrave Macmillan, 2014), 228–40.

15. Burke's list comes from his description of the topics associated with science at the expense of being associated with human relationships. Toolan, *Narrative,* 8; Kenneth Burke, "Four Master Tropes," *Kenyon Review* 3, no. 4 (1941): 423, 424.

16. Shawn Wilson, *Research Is Ceremony: Indigenous Research Methods* (Halifax, NS: Fernwood, 2008).

17. Kenneth Burke, *A Grammar of Motives* (Berkeley: University of California Press, 1945), 171.

18. For example, studies have found that identification with characters improved narrative persuasion, enjoyment of the narrative, the complexity of reflection on the story, and a story's influence on attitudes and beliefs. Hans Hoeken, Matthijs Kolthoff, and José Sanders, "Story Perspective and Character Similarity as Drivers of Identification and Narrative Persuasion: Perspective, Similarity, and Identification," *Human Communication Research* 42, no. 2 (2016): 292–311, https://doi.org/10.1111/hcre.12076; Michael D. Slater and Donna Rouner, "Entertainment-Education and Elaboration Likelihood: Understanding the Processing of Narrative Persuasion," *Communication Theory* 12, no. 2 (2002): 173–91, https://doi.org/10.1111/j.1468-2885.2002.tb00265.x; Juan-José Igartua, "Identification with Characters and Narrative Persuasion through Fictional Feature Films," *Communications* 35, no. 4 (2010): 347–73, https://doi.org/10.1515/comm.2010.019.

19. H. Porter Abbott, *The Cambridge Introduction to Narrative* (Cambridge: Cambridge University Press, 2008), 1.

20. Michael F. Dahlstrom and Shirley S. Ho, "Ethical Considerations of Using Narrative to Communicate Science," *Science Communication* 34, no. 5 (2012): 593, https://doi.org/10.1177/1075547012454597; see also Marie-Laure Ryan, "Toward a Definition of Narrative," in *The Cambridge Companion to Narrative,* ed. David Herman (Cambridge: Cambridge University Press, 2007), 22–35.

21. Burke, *A Grammar of Motives,* 187, emphasis removed.

22. Sang-Kyong Lee, "Edward Gordon Craig and Japanese Theatre," *Asian Theatre Journal,* 2000, 229, https://www.jstor.org/stable/1124490.

23. Rob Nixon, *Slow Violence and the Environmentalism of the Poor* (Cambridge, MA: Harvard University Press, 2011), 2.

24. Nixon.

25. Leah Ceccarelli, "CRISPR as Agent: A Metaphor That Rhetorically Inhibits the Prospects for Responsible Research," *Life Sciences, Society and Policy* 14, no. 1 (2018): 10, https://doi.org/10.1186/s40504-018-0088-8.

26. Lauren R. Kolodziejski, "Harms of Hedging in Scientific Discourse: Andrew Wakefield and the Origins of the Autism Vaccine Controversy," *Technical Communication Quarterly* 23, no. 3 (2014): 171, https://doi.org/10.1080/10572252.2013.816487.

27. Ceccarelli, "CRISPR as Agent," 11.

28. Robert K. Merton, *The Sociology of Science: Theoretical and Empirical Investigations* (Chicago: University of Chicago Press, 1979).

29. Robert Merton, cited in Richard Wunderlich, "The Scientific Ethos: A Clarification," *British Journal of Sociology* 25, no. 3 (1974): 375, https://doi.org/10.2307/589402.

30. Wunderlich, 376.

31. Intergovernmental Panel on Climate Change, *Summary for Policymakers—AR5 Synthesis Report: Climate Change 2014*, 2, https://www.ipcc.ch/report/ar5/syr/.

32. Emma Frances Bloomfield and Chris Manktelow, "Climate Communication and Storytelling," *Climatic Change* 167, no. 34 (2021): 3, https://doi.org/10.1007/s10584-021-03199-6.

33. Nature Conservancy, "Climate Change: Frequently Asked Questions," last updated October 7, 2021, https://www.nature.org/en-us/what-we-do/our-priorities/tackle-climate-change/climate-change-stories/climate-change-frequently-asked-questions/.

34. Royal Society, "A Short Guide to Climate Science," March 2020, https://royalsociety.org/-/media/policy/projects/climate-evidence-causes/climate-change-q-and-a.pdf.

35. American Association for the Advancement of Science, "AAAS Reaffirms Statements on Climate Change and Integrity," December 4, 2009, https://www.aaas.org/news/aaas-reaffirms-statements-climate-change-and-integrity.

36. NASA Global Climate Change, "Scientific Consensus: Earth's Climate Is Warming," last updated September 1, 2022, https://climate.nasa.gov/scientific-consensus.

37. Donald R. Prothero, *Evolution: What the Fossils Say and Why It Matters* (New York: Columbia University Press, 2007), 101, 102.

38. Burke, *A Grammar of Motives.*

39. Understanding Evolution, *Misconceptions about Evolution* (2012), 4, 5, https://evolution.berkeley.edu/evolibrary/misconceptions_about_evolution.pdf.

40. Cameron McPherson Smith, *The Fact of Evolution* (Amherst, NY: Prometheus Books, 2011), 180, emphasis added.

41. John Angus Campbell, "Darwin and the Origin of Species: The Rhetorical Ancestry of an Idea," *Speech Monographs* 37, no. 1 (1970): 9, https://doi.org/10.1080/03637757009375644.

42. Charles Darwin, *On the Origin of Species,* anniv. ed. (1859; repr.; New York: Signet, 2003), 53.

43. Understanding Evolution, *Misconceptions about Evolution,* 5; Prothero, *Evolution,* 101.

44. Prothero, *Evolution,* 101.

45. National Academy of Sciences, *Science, Evolution, and Creationism* (Washington, DC: National Academies Press, 2008), 21, https://doi.org/10.17226/11876.

46. National Academy of Sciences, *Science and Creationism: A View from the National Academy of Sciences,* 2nd ed. (Washington, DC: National Academies Press, 1999), 3, https://doi.org/10.17226/6024.

47. National Academy of Sciences, *Science, Evolution, and Creationism,* 17.

48. American Academy of Family Physicians, "Childhood Vaccines: What They Are and Why Your Child Needs Them," *Family Doctor,* March 18, 2020, https://familydoctor.org/childhood-vaccines-what-they-are-and-why-your-child-needs-them/.

49. Centers for Disease Control, "Vaccine Information Statement—Live Intranasal Influenza," October 22, 2019, https://www.cdc.gov/vaccines/hcp/vis/vis-statements/flulive.html, emphasis added.

50. I use the term *instructional communication* to indicate communication given to instruct and educate. This usage should not be confused with the term's use in communication pedagogy, which analyzes the communicative factors that influence classroom learning.

51. In citing this work, I adopt the language of narrative and nonnarrative used by the authors. To phrase this study in narrative web terminology, I would say that the authors presented two narratives regarding cervical cancer where one was stripped of many of its micro-ring narrative elements, such as relatable characters, a clear plot, and a narrowed scope. Sheila T. Murphy et al., "Narrative versus Nonnarrative: The Role of Identification, Transportation, and Emotion in Reducing Health Disparities," *Journal of Communication* 63, no. 1 (2013): 116–37, https://doi.org/10.1111/jcom.12007.

52. Murphy et al., 116–37.

53. Flor M. Muñoz, "Which Flu Vaccine Should My Children Get This Year?," Healthy Children, November 11, 2019, https://www.healthychildren.org/English/safety-prevention/immunizations/Pages/Prepare-Your-Family-for-Flu-Season.aspx.

54. World Health Organization, "COVID-19 Vaccines," accessed August 30, 2022, https://www.who.int/emergencies/diseases/novel-coronavirus-2019/covid-19-vaccines.

55. Centers for Disease Control, "Vaccine Effectiveness Studies," accessed August 30, 2022, https://www.cdc.gov/coronavirus/2019-ncov/vaccines/effectiveness/how-they-work.html.

56. Centers for Disease Control, "Vaccine Effectiveness Studies."

57. Milstein, "Nature Identification," 3–24; Emma Frances Bloomfield, "Transcorporeal Identification and Strategic Essentialism in Eco-Horror: *mother!*'s Ecofeminist Rhetorical Strategies," *Environmental Communication* 15, no. 3 (2021): 339–52, https://doi.org/10.1080/17524032.2020.1833059.

58. Sara J. ElShafie, "Making Science Meaningful for Broad Audiences through Stories," *Integrative and Comparative Biology* 58, no. 6 (2018): 1213–23, https://doi.org/10.1093/icb/icy103.

59. ElShafie, 1217.

60. The term *pathetic fallacy* is often attributed to John Ruskin, who defined it as ascribing human feelings to inanimate objects. It is considered a fallacy because the ascription of emotions is understood to be inaccurate, thereby revealing a Western bias against objects and nonhuman life as "non-feeling." John Ruskin, *Modern Painters*, vol. 3, https://www.gutenberg.org/cache/epub/38923/pg38923-images.html#CHAPTER_XII.

61. Leanne Betasamosake Simpson, *As We Have Always Done: Indigenous Freedom through Radical Resistance* (Minneapolis: University of Minnesota Press, 2017); Daniel Wildcat, *Red Alert!: Saving the Planet with Indigenous Knowledge* (Wheat Ridge, CO: Fulcrum, 2009); Robin Wall Kimmerer, *Braiding Sweetgrass: Indigenous Wisdom, Scientific Knowledge and the Teachings of Plants* (Minneapolis: Milkweed Editions, 2015); Kunisue, "Listening to the Natural World," 228–40.

62. María Inés Palleiro, "Animals, Tale Types, and Belief Narratives in Argentinean Folklore," *Folklore: Electronic Journal of Folklore* 77 (2019): 13, https://doi.org/10.7592/FEJF2019.77.palleiro.

63. Palleiro, 11–38.

64. P. Sol Hart and Erik C. Nisbet, "Boomerang Effects in Science Communication: How Motivated Reasoning and Identity Cues Amplify Opinion Polarization about Climate Mitigation Policies," *Communication Research* 39, no. 6 (2012): 701–23, https://doi.org/10.1177/0093650211416646.

65. Megan R. Dillow and Keith Weber, "An Experimental Investigation of Social Identification on College Student Organ Donor Decisions," *Communication Research Reports* 33, no. 3 (2016): 239–46, https://doi.org/10.1080/08824096.2016.1186630.

66. Emma Frances Bloomfield et al., "The Effects of Establishing Intimacy and Consubstantiality on Group Discussions about Climate Change Solutions," *Science Communication* 42, no. 3 (2020): 369–94, https://doi.org/10.1177/1075547020927017; Lyn M. van Swol et al., "Fostering Climate Change Consensus: The Role of Intimacy in Group Discussions," *Public Understanding of Science* 31, no. 1 (2022), 103–18, https://doi.org/10.1177/09636625211020661.

67. Burke, *A Grammar of Motives*, xix.

68. American Physical Society, "15.3 Statement on Earth's Changing Climate," November 14, 2015, http://www.aps.org/policy/statements/15_3.cfm.

69. American Meteorological Society, "Climate Change: An Information Statement," April 15, 2019, https://www.ametsoc.org/index.cfm/ams/about-ams/ams-statements/statements-of-the-ams-in-force/climate-change1/.

70. NASA Global Climate Change, "The Causes of Climate Change," accessed June 8, 2020, https://climate.nasa.gov/causes.

71. Environmental Protection Agency, "Causes of Climate Change," Overviews and Factsheets, January 19, 2017, https://www.epa.gov/climatechange-science/causes-climate-change.

72. Environmental Protection Agency, "Future of Climate Change," Overviews and Factsheets, January 19, 2017, https://climatechange.chicago.gov/climate-change-science/future-climate-change.

73. Environmental Protection Agency, "Climate Change Indicators in the United States," accessed June 8, 2020, https://www.epa.gov/climate-indicators.

74. Environmental Protection Agency, "Causes of Climate Change."

75. American Meteorological Society, "Climate Change: An Information Statement," emphasis added.

76. NASA Global Climate Change, "The Causes of Climate Change."

77. National Academy of Sciences, *Climate Change: Evidence and Causes* (Washington, DC: National Academies Press, 2014), https://doi.org/10.17226/18730.

78. Environmental Protection Agency, "Climate Change Indicators in the United States."

79. Environmental Protection Agency, "Climate Change Indicators in the United States."

80. Intergovernmental Panel on Climate Change, *AR5 Synthesis Report: Climate Change 2014*, https://www.ipcc.ch/report/ar5/syr/.

81. American Meteorological Society, "Climate Change: An Information Statement."

82. Emma Frances Bloomfield, "The Reworking of Evangelical Christian Ecocultural Identity in the Creation Care Movement," in *The Routledge Handbook of Ecocultural Identity*, ed. Tema Milstein and José Castro-Sotomayor (New York: Routledge, 2020), 201.

83. Intergovernmental Panel on Climate Change, *AR5 Synthesis Report*.

84. U.S. Global Change Research Program, "Impacts on Society," *GlobalChange*, accessed January 2, 2019, https://www.globalchange.gov/climate-change/impacts-society.

85. Cornwall Alliance, *A Renewed Call to Truth, Prudence, and Protection of the Poor: An Evangelical Examination of the Theology, Science, and Economics of Global Warming*, 2009, http://www.cornwallalliance.org/wp-content/uploads/2014/04/a-renewed-call-to-truth-prudence-and-protection-of-the-poor.pdf.

86. Brendan O'Connor, "How Fossil Fuel Money Made Climate Change Denial the Word of God," *Splinter*, August 8, 2017, https://splinternews.com/how-fossil-fuel-money-made-climate-denial-the-word-of-g-1797466298.

87. Donn Dears, *Nothing to Fear: A Bright Future for Fossil Fuels* (Arlington Heights, IL: Critical Thinking Press, 2015), http://store.heartland.org/shop/nothing-to-fear-a-bright-future-for-fossil-fuels/.

88. Bruno Latour, "Agency at the Time of the Anthropocene," *New Literary History* 45, no. 1 (2014): 1, https://doi.org/10.1353/nlh.2014.0003.

89. Anthony Leiserowitz, "Climate Change Risk Perception and Policy Preferences: The Role of Affect, Imagery, and Values," *Climatic Change* 77, no. 1 (2006): 56, https://doi.org/10.1007/s10584-006-9059-9.

90. Alfred Tennyson, "In Memoriam A. H. H.," 1850 [2013], http://www.online-literature.com/donne/718/.

91. Answers in Genesis, "If God Doesn't Matter to Him, Do You?," accessed September 6, 2022, https://answersingenesis.org/morality/you-matter-to-god/.

92. Ken Ham, "Violence and Evolution?," Answers in Genesis, February 3, 2022, https://answersingenesis.org/media/audio/answers-with-ken-ham/volume-144/violence-and-evolution/.

93. Heidi Yoston Lawrence, *Vaccine Rhetorics* (Columbus: Ohio State University Press, 2020); Jennifer A. Reich, *Calling the Shots: Why Parents Reject Vaccines* (New York: NYU Press, 2018).

94. Centers for Disease Control, "Making the Vaccine Decision: Common Concerns," June 16, 2020, https://www.cdc.gov/vaccines/parents/why-vaccinate/vaccine-decision.html.

95. Centers for Disease Control, "Making the Vaccine Decision."

96. Jeanne P. Spencer, Ruth H. Trondsen Pawlowski, and Stephanie Thomas, "Vaccine Adverse Events: Separating Myth from Reality," *American Family Physician* 95, no. 12 (2017): 791, https://www.aafp.org/pubs/afp/issues/2017/0615/p786.html.

97. American Academy of Pediatrics, *Vaccine Advocacy Snapshot*, 2020, https://downloads.aap.org/DOFA/AAPVaccinationsOnePager.pdf.

98. Spencer, Trondsen Pawlowski, and Thomas, "Vaccine Adverse Events," 787.

99. Spencer, Trondsen Pawlowski, and Thomas, 791.

100. Vaccinate Your Family, "Babies & Children," June 24, 2020, https://www.vaccinateyourfamily.org/which-vaccines-does-my-family-need/babies-children/.

101. Hoppin applied Fisher's narrative paradigm to vaccine hesitant discourse and interpreted those stories as having high coherence and fidelity for parents searching for answers regarding their children's autism diagnoses. Shari Hoppin, "Applying the Narrative Paradigm to the Vaccine Debates," *American Communication Journal* 18, no. 2 (2016): 50.

102. Hoppin, 52.

103. Charles Piller, "'This Is Insane!' Many Scientists Lament Trump's Embrace of Risky Malaria Drugs for Coronavirus," *Science*, March 26, 2020, https://www.science.org/content/article/insane-many-scientists-lament-trump-s-embrace-risky-malaria-drugs-coronavirus.

104. Jeff Tollefson, "How Trump Damaged Science—and Why It Could Take Decades to Recover," *Nature* 586, no. 7828 (2020): 190, https://www.nature.com/articles/d41586-020-02800-9.

105. Alberto Giubilini and Julian Savulescu, "Shaming Unvaccinated People Has to Stop. We've Turned into an Angry Mob and It's Getting Ugly," *The Conversation,* December 15, 2021, http://theconversation.com/shaming-unvaccinated-people-has-to-stop-weve-turned-into-an-angry-mob-and-its-getting-ugly-173137.

106. Marc Fisher, "Fed Up with Anti-Maskers, Mask Advocates Are Demanding Mandates, Fines—and Common Courtesy," *Washington Post,* August 18, 2020, https://www.washingtonpost.com/politics/anti-mask-outrage/2020/08/18/40d0bc96-d746-11ea-930e-d88518c57dcc_story.html.

107. There have been numerous incidents of COVID-19–related violence that I could reference. Violence was especially common in 2020 and 2021 when lockdowns and mask mandates were broadly in effect. Since vaccines have become widespread and Centers for Disease Control guidance has changed, there is no longer a reliable correlation between being unmasked and being COVID-19 skeptical. Jennifer Sinco Kelleher, Terry Tang, and Olga R. Rodriguez, "Mask, Vaccine Conflicts Descend into Violence and Harassment," *AP NEWS,* August 21, 2021, https://apnews.com/article/health-coronavirus-pandemic-2eba81e-be3bd54b3bcde890b8cf11c70; Fisher, "Fed Up with Anti-Maskers."

108. Stephen J. Flusberg, Teenie Matlock, and Paul H. Thibodeau, "Metaphors for the War (or Race) against Climate Change," *Environmental Communication* 11, no. 6 (2017): 769–83, https://doi.org/10.1080/17524032.2017.1289111; Steven Schwarze, "Environmental Melodrama," *Quarterly Journal of Speech* 92, no. 3 (2006): 239–61, https://doi.org/10.1080/00335630600938609.

109. Kate Yoder, "Is Waging 'War' the Only Way to Take on the Coronavirus?," *Grist,* April 15, 2020, https://grist.org/climate/is-waging-war-the-only-way-to-take-on-the-coronavirus/.

110. Associated Press, "More than 9,000 Anti-Asian Incidents Have Been Reported Since the Pandemic Began," *NPR,* August 12, 2021, https://www.npr.org/2021/08/12/1027236499/anti-asian-hate-crimes-assaults-pandemic-incidents-aapi.

111. Emma Frances Bloomfield, *Communication Strategies for Engaging Climate Skeptics: Religion and the Environment* (New York: Routledge, 2019).

112. Burke, *A Grammar of Motives,* xxi.

113. Burke, xxi.

114. Judith Butler, *Gender Trouble: Feminism and the Subversion of Identity* (New York: Routledge, 2006), 142.

115. Butler, 142.

116. Stuart Hall, "Who Needs Identity?," in *Questions of Cultural Identity,* ed. Stuart Hall and Paul du Gay (Los Angeles: SAGE, 1996), 5.

117. Julia Sangervo, Kirsti M. Jylhä, and Panu Pihkala, "Climate Anxiety: Conceptual Considerations, and Connections with Climate Hope and Action," *Global Environmental Change* 76 (2022): 2, https://doi.org/10.1016/j.gloenvcha.2022.102569.

118. Sangervo, Jylhä, and Pihkala, "Climate Anxiety," 1–11; Lorraine Whitmarsh et al., "Climate Anxiety: What Predicts It and How Is It Related to Climate Action?," *Journal of Environmental Psychology* 83 (2022): 1–10, https://doi.org/10.1016/j.jenvp.2022.101866; Britt Wray, *Generation Dread: Finding Purpose in an Age of Climate Crisis* (Toronto: Knopf Canada, 2022).

119. Mark Kaufman, "The Devious Fossil Fuel Propaganda We All Use," *Mashable,* July 13, 2020, https://mashable.com/feature/carbon-footprint-pr-campaign-sham.

120. ClientEarth, "Fossil Fuels and Climate Change: The Facts," February 18, 2022,https://www.clientearth.org/latest/latest-updates/stories/fossil-fuels-and-climate-change-the-facts/.

121. Amy Westervelt, "Big Oil Is Trying to Make Climate Change Your Problem to Solve. Don't Let Them," *Rolling Stone,* May 14, 2021, https://www.rollingstone.com/politics/politics-news/climate-change-exxonmobil-harvard-study-1169682/.

122. Marcelle McManus, "The 'Carbon Footprint' Was Co-Opted by Fossil Fuel Companies to Shift Climate Blame—Here's How It Can Serve Us Again," *The Conversation,* May 27, 2022, http://theconversation.com/the-carbon-footprint-was-co-opted-by-fossil-fuel-companies-to-shift-climate-blame-heres-how-it-can-serve-us-again-183566.

123. Jay Michaelson, "Why Your Carbon Footprint Is Meaningless," *Daily Beast,* September 16, 2019, https://www.thedailybeast.com/why-your-carbon-footprint-is-meaningless.

124. Burke, *A Grammar of Motives,* 79.

125. Understanding Evolution, *Misconceptions about Evolution,* 2.

126. John Archibald, *One Plus One Equals One: Symbiosis and the Evolution of Complex Life* (New York: Oxford University Press, 2014), 13.

127. It is important to note that the use of the term *blind* is ableist in this context in connoting that a lack of eyesight means no critical thinking. Richard Dawkins, *The Blind Watchmaker: Why the Evidence of Evolution Reveals a Universe without Design* (New York: W. W. Norton, 1996), 42.

128. Dawkins, 43.

129. Smith, *The Fact of Evolution,* 181.

130. Stephen Meyer and Keith Fox, "Signature in the Cell—Stephen C. Meyer vs. Keith Fox," Unbelievable?, November 2011, audio, http://www.premierradio.org.uk/listen/ondemand.aspx?mediaid=%7BD5D3E5D1-697C-4348-87E6-7B6EED18E0AC%7D.

131. Understanding Evolution, *Misconceptions about Evolution,* 5, 3.

132. Smithsonian National Museum of Natural History, "Frequently Asked Questions," Smithsonian Institution's Human Origins Program, December 22, 2009, http://humanorigins.si.edu/education/frequently-asked-questions.

133. NASA, "NASA Research Reveals Major Insight into Evolution of Life on Earth," August 19, 2009, https://www.nasa.gov/topics/earth/features/astrobiology.html.

134. Mark A. Largent, *Vaccine: The Debate in Modern America* (Baltimore: Johns Hopkins University Press, 2012), 158.

135. Family Doctor, "The Importance of Vaccinations," last updated April 2022, https://familydoctor.org/the-importance-of-vaccinations/; Muñoz, "Which Flu Vaccine Should My Children Get This Year?"

136. Stephanie Wasserman, "Immunizations and Vaccines for Children (VFC) Update," accessed August 30, 2022, https://slideplayer.com/slide/14472889/.

137. Stephanie Willes, "The Affordances of the Internet and the Creation of Villains and Heroes in Anti-Vaccine Stories: A Narrative Criticism of VaxTruth" (master's thesis, University of Nevada, Las Vegas, 2019).

138. Marcella Piper-Terry, "Why I Don't Vaccinate," *Vaxxed: Vaccine-Injury*, accessed October 3, 2022, https://vaccine-injury.info/marcella-piper-terry-why-i-do-not-vaccinate; Willes, "The Affordances of the Internet."

139. Piper-Terry, "Why I Don't Vaccinate."

140. Walter R. Fisher, "Narration as a Human Communication Paradigm: The Case of Public Moral Argument," *Communication Monographs* 51, no. 1 (1984): 14, https://doi.org/10.1080/03637758409390180.

CHAPTER 3

1. Chloe Powell, "Cli-Fi Cinema: An Epideictic Rhetoric of Blame" (master's thesis, University of Nevada, Las Vegas, 2017), https://digitalscholarship.unlv.edu/thesesdissertations/3026; E. Ann Kaplan, *Climate Trauma: Foreseeing the Future in Dystopian Film and Fiction* (New Brunswick, NJ: Rutgers University Press, 2015); Michael Svoboda, "Cli-Fi on the Screen(s): Patterns in the Representations of Climate Change in Fictional Films," *Wiley Interdisciplinary Reviews: Climate Change* 7, no. 1 (2016): 43–64, https://doi.org/10.1002/wcc.381; Emma Frances Bloomfield, "Transcorporeal Identification and Strategic Essentialism in Eco-Horror: *mother!*'s Ecofeminist Rhetorical Strategies," *Environmental Communication* 15, no. 3 (2021): 339–52, https://doi.org/10.1080/17524032.2020.1833059.

2. Emma Frances Bloomfield, "*mother!* and the Horror of Environmental Abuse," in *The Politics of Horror*, ed. Damien Picariello (Cham, Switzerland: Palgrave Macmillan, 2020), 187–98.

3. Timothy Morton, *Hyperobjects: Philosophy and Ecology after the End of the World* (Minneapolis: University of Minnesota Press, 2013).

4. Fox proposed that climate communicators should use terms such as *climate fires* or *climate-driven fires* to describe wildfires as opposed to *natural*

disasters in order to "shape public discourse by connecting climate issues to wildland fire each time we talk about them." Rebekah L. Fox, "How We Talk about Fire Matters," *ECD NCA* (blog), August 17, 2022, https://sites.google.com/view/ecdnca/blog#h.bsxiwdok63zo.

5. "'Never Seen Climate Carnage' Like Pakistan Floods, Says UN Chief," *Al Jazeera,* September 10, 2022, https://www.aljazeera.com/news/2022/9/10/never-seen-climate-carnage-like-in-pakistan-un-chief.

6. Joshua DiCaglio, *Scale Theory: A Nondisciplinary Inquiry* (Minneapolis: University of Minnesota Press, 2021), 2.

7. Chris Ingraham, "The Scope and Autonomy of Personal Narrative," *Written Communication* 34, no. 1 (2017): 61, https://doi.org/10.1177/0741088316683147.

8. Walter R. Fisher, "Narration as a Human Communication Paradigm: The Case of Public Moral Argument," *Communication Monographs* 51, no. 1 (1984): 1–22, https://doi.org/10.1080/03637758409390180; Marie-Laure Ryan, "Toward a Definition of Narrative," in *The Cambridge Companion to Narrative,* ed. David Herman (Cambridge: Cambridge University Press, 2007), 22–35.

9. Nam Wook Kim et al., "Visualizing Nonlinear Narratives with Story Curves," *IEEE Transactions on Visualization and Computer Graphics* 24, no. 1 (2018): 595, https://doi.org/10.1109/TVCG.2017.2744118.

10. Strawson argues that stories are not a "random or radically unconnected sequence of events." Galen Strawson, "Against Narrativity," *Ratio* 17, no. 4 (2004): 439, https://doi.org/10.1111/j.1467-9329.2004.00264.x.

11. Brian Richardson, "Recent Concepts of Narrative and the Narratives of Narrative Theory," *Style* 34, no. 2 (2000): 169–70, https://www.jstor.org/stable/10.5325/style.34.2.168.

12. Kenneth Burke, *Counter-Statement* (Berkeley: University of California Press, 1968), 124.

13. The phrase "canons of rhetoric" refers to five components of building speeches in classical Greek and Roman orations. In addition to arrangement, they include invention (or the creation of ideas and arguments), style, memory, and delivery.

14. Wilhelm Wuellner, "Arrangement," in *Handbook of Classical Rhetoric in the Hellenistic Period: 330 B.C.–A.D. 400,* ed. Stanley E. Porter (Leiden, Netherlands: Brill, 2001), 51.

15. Rhetorical scholars may also call these repeated forms "genres," which constitute categories of discourse based on stylistic, substantive, and situational similarities. Lewis B. Hershey, "Burke's Aristotelianism: Burke and Aristotle on Form," *Rhetoric Society Quarterly* 16, no. 3 (1986): 184, https://doi.org/10.1080/02773948609390748; Karlyn Kohrs Campbell and Kathleen Hall Jamieson, "Form and Genre in Rhetorical Criticism: An Introduction," in *Form and Genre: Shaping Rhetorical Action,* ed. Karlyn Kohrs Campbell and Kathleen Hall Jamieson (Falls Church, VA: Speech Communication Association, 1978), 9–32, https://eric.ed.gov/?id=ED151893.

16. Susanne K. Langer, *Mind: An Essay on Human Feeling* (Baltimore: Johns Hopkins University Press, 1988).

17. Burke, *Counter-Statement*, xi.

18. Wuellner, "Arrangement," 55.

19. Phaedra C. Pezzullo, "Performing Critical Interruptions: Stories, Rhetorical Invention, and the Environmental Justice Movement," *Western Journal of Communication* 65, no. 1 (2001): 1–25, https://doi.org/10.1080/10570310109374689.

20. The Warren County PCB landfill was eventually cleaned up in 2004, three years after Pezzullo's article was published and more than two decades after the landfill was created in 1982. Pezzullo, "Performing Critical Interruptions," 2.

21. Randall A. Lake, "Between Myth and History: Enacting Time in Native American Protest Rhetoric," *Quarterly Journal of Speech* 77, no. 2 (1991): 123–51, https://doi.org/10.1080/00335639109383949; Leanne Betasamosake Simpson, *As We Have Always Done: Indigenous Freedom through Radical Resistance* (Minneapolis: University of Minnesota Press, 2017); Jan Hare, "'They Tell a Story and There's Meaning Behind That Story': Indigenous Knowledge and Young Indigenous Children's Literacy Learning," *Journal of Early Childhood Literacy* 12, no. 4 (2012): 389–414, https://doi.org/10.1177/1468798411417378.

22. Lake stresses the importance of evaluating Indigenous rhetoric on its own terms and not whether it complies with Western and non-Indigenous rhetorical standards. Randall A. Lake, "Enacting Red Power: The Consummatory Function in Native American Protest Rhetoric," *Quarterly Journal of Speech* 69, no. 2 (1983): 127–42, https://doi.org/10.1080/00335638309383642.

23. Kenneth Burke, *A Grammar of Motives* (Berkeley: University of California Press, 1945), 77.

24. Jerome S. Bruner, "Life as Narrative," *Social Research* 54, no. 1 (1987): 17, https://www.jstor.org/stable/40970444.

25. José Castro-Sotomayor, "Emplacing Climate Change: Civic Action at the Margins," *Frontiers in Communication* 4 (2019): 8, https://www.frontiersin.org/article/10.3389/fcomm.2019.00033.

26. Rainer Romero-Canyas et al., "Bringing the Heat Home: Television Spots about Local Impacts Reduce Global Warming Denialism," *Environmental Communication* 13, no. 6 (2019): 740–60, https://doi.org/10.1080/17524032.2018.1455725.

27. Shawn Wilson, *Research Is Ceremony: Indigenous Research Methods* (Halifax, NS: Fernwood, 2008), 87.

28. Misia Landau, *Narratives of Human Evolution* (New Haven, CT: Yale University Press, 1993), ix.

29. Kunisue's research was based on local traditional knowledge on the Big Island of Hawai'i (also called Hawai'i Island). Yukari Kunisue, "Listening to the Natural World: Ecopsychology of Listening from a Hawai'ian Spiritual Perspective," in *Voice and Environmental Communication*, ed. Jennifer Peeples and

Stephen Depoe (New York: Palgrave Macmillan, 2014), 228–40, https://doi.org/10.1057/9781137433749_12.

30. Meera Subramanian, "Anthropocene Now: Influential Panel Votes to Recognize Earth's New Epoch," *Nature* (2019), https://doi.org/10.1038/d41586-019-01641-5.

31. EarthHow, "What Is Earth's Geological Time Scale?," July 25, 2017, https://earthhow.com/earth-geological-time-scale/, emphasis in original.

32. National Academy of Sciences and the Royal Society, "Climate Change: Evidence and Causes," February 26, 2014, https://doi.org/10.17226/18730.

33. American Meteorological Society, "Climate Change: An Information Statement," April 15, 2019, https://www.ametsoc.org/index.cfm/ams/about-ams/ams-statements/statements-of-the-ams-in-force/climate-change1/.

34. Alistair Hunt and Tim Taylor, "Values and Cost–Benefit Analysis: Economic Efficiency Criteria in Adaptation," in *Adapting to Climate Change: Thresholds, Values, Governance,* ed. Irene Lorenzoni, Karen L. O'Brien, and W. Neil Adger (Cambridge: Cambridge University Press, 2009), 197–211; Sabine Pahl et al., "Perceptions of Time in Relation to Climate Change," *WIREs Climate Change* 5, no. 3 (2014): 375–88, https://doi.org/10.1002/wcc.272; Bruce Tonn, Angela Hemrick, and Fred Conrad, "Cognitive Representations of the Future: Survey Results," *Futures* 38, no. 7 (2006): 810–29, https://doi.org/10.1016/j.futures.2005.12.005.

35. Peter Rudiak-Gould, "'We Have Seen It with Our Own Eyes': Why We Disagree about Climate Change Visibility," *Weather, Climate, and Society* 5, no. 2 (2013): 121, https://doi.org/10.1175/WCAS-D-12-00034.1.

36. Pew Research Center, "4. Important Issues in the 2020 Election," *U.S. Politics & Policy* (blog), August 13, 2020, https://www.pewresearch.org/politics/2020/08/13/important-issues-in-the-2020-election/.

37. Pahl et al., "Perceptions of Time in Relation to Climate Change," 376, emphasis added.

38. Fisher, "Narration as a Human Communication Paradigm," 2.

39. National Academy of Sciences, *Science and Creationism: A View from the National Academy of Sciences,* 2nd ed. (Washington, DC: National Academies Press, 1999), 3, https://doi.org/10.17226/6024.

40. Southern California Seminary, "SCS Fall 2013 Debate: Did God Use Evolution?," October 24, 2013, https://vimeo.com/77736640.

41. National Research Council, *Thinking Evolutionarily: Evolution Education across the Life Sciences—Summary of a Convocation* (Washington, DC: National Academies Press, 2012), 44, 25.

42. National Research Council, 36.

43. Ker Than, "All Species Evolved from Single Cell, Study Finds," *National Geographic,* May 14, 2010, https://www.nationalgeographic.com/news/2010/5/100513-science-evolution-darwin-single-ancestor/.

44. Wayne Spencer, "Why Recent Creation?," Answers in Genesis, June 16, 2009, https://answersingenesis.org/why-does-creation-matter/why-recent-creation/.

45. Rudiak-Gould, "'We Have Seen It with Our Own Eyes,'" 123.

46. Rudiak-Gould, 123.

47. U.S. Global Change Research Program, *Fourth National Climate Assessment,* 2018, https://nca2018.globalchange.gov; Intergovernmental Panel on Climate Change, *AR5 Synthesis Report: Climate Change 2014,* https://www.ipcc.ch/report/ar5/syr/.

48. Intergovernmental Panel on Climate Change.

49. Intergovernmental Panel on Climate Change.

50. P. Sol Hart and Erik C. Nisbet, "Boomerang Effects in Science Communication: How Motivated Reasoning and Identity Cues Amplify Opinion Polarization about Climate Mitigation Policies," *Communication Research* 39, no. 6 (2012), 701–23, https://doi.org/10.1177/0093650211416646.

51. Dylan M. Harris, "Telling the Story of Climate Change: Geologic Imagination, Praxis, and Policy," *Energy Research & Social Science* 31 (2017): 182, https://doi.org/10.1016/j.erss.2017.05.027.

52. American Academy of Pediatrics, *Vaccine Advocacy Snapshot,* 2020, https://downloads.aap.org/DOFA/AAPVaccinationsOnePager.pdf.

53. American Academy of Pediatrics.

54. Centers for Disease Control, "Vaccine Information Statement—Live Intranasal Influenza," October 22, 2019, https://www.cdc.gov/vaccines/hcp/vis/vis-statements/flulive.html.

55. World Health Organization, "Welcome—WHO Vaccine Safety Basics," 2013, https://vaccine-safety-training.org/.

56. Marcella Piper-Terry, "Vaccinations: Your Child vs. 'The Greater Good,'" VaxTruth, July 5, 2015, http://vaxtruth.org/2015/07/vaccinesyourchild2015/.

57. Centers for Disease Control, "Six Things You Need to Know about Vaccines," January 14, 2020, https://www.cdc.gov/vaccines/vac-gen/vaxwithme.html.

58. Family Doctor, "The Importance of Vaccinations," last updated April 2022, https://familydoctor.org/the-importance-of-vaccinations/.

59. Joshua A. Krisch, "What Is Herd Immunity?," *Live Science,* July 13, 2020, https://www.livescience.com/herd-immunity.html.

60. Marcella Piper-Terry, "Vaccines, Autism, and Susceptible Groups: Is Your Child at Increased Risk?," VaxTruth, September 10, 2014, http://vaxtruth.org/2014/09/susceptible-groups/.

61. Marcella Piper-Terry, "Meet Jordan," VaxTruth, July 19, 2015, http://vaxtruth.org/2015/07/meet-jordan/.

62. Emma Frances Bloomfield, *Communication Strategies for Engaging Climate Skeptics: Religion and the Environment* (New York: Routledge, 2019).

63. Fahnestock included "the wonder" and "the application" under the concept of "scientific accommodation," which is the communication of technical, scientific topics to lay audiences. Appeals to wonder emphasize the "never before" made achievements or the "amazing powers and secrets" of a scientific breakthrough. Appeals to application emphasize direct benefits extending from the scientific advancement, such as technological "spin-offs" from investing in space exploration. Appeals to application will be discussed further related to the content wedge. Jeanne Fahnestock, "Accommodating Science: The Rhetorical Life of Scientific Facts," *Written Communication* 3, no. 3 (1986): 279, https://doi.org/10.1177/0741088386003003001.

64. Rudiak-Gould, "'We Have Seen It with Our Own Eyes,'" 123.

65. World Health Organization, "Welcome—WHO Vaccine Safety Basics."

66. Pahl et al., "Perceptions of Time in Relation to Climate Change," 376.

67. DiCaglio, *Scale Theory*.

68. Rudiak-Gould, "'We Have Seen It with Our Own Eyes,'" 122.

69. NASA Global Climate Change, "Climate Change Evidence: How Do We Know?," accessed June 8, 2020, https://climate.nasa.gov/evidence.

70. National Academy of Sciences and the Royal Society, *Climate Change: Evidence and Causes*, 2014, 3, https://doi.org/10.17226/18730.

71. Current Results, "Average Annual Temperatures for Large US Cities," accessed June 22, 2020, https://www.currentresults.com/Weather/US/average-annual-temperatures-large-cities.php.

72. American Meteorological Society, "Climate Change: An Information Statement."

73. Environmental Protection Agency, "Climate Change Indicators in the United States," accessed June 8, 2020, https://www.epa.gov/climate-indicators.

74. Meghan Keneally, "Trump's Global Warming Tweet Is a 'Troll Job,' Not Based in Science, Experts Say," *ABC News*, December 29, 2017, https://abcnews.go.com/US/trumps-global-warming-tweet-troll-job-based-science/story?id=52044473.

75. Rudiak-Gould, "'We Have Seen It with Our Own Eyes,'" 122.

76. Rudiak-Gould, "'We Have Seen It with Our Own Eyes,'" 120–21.

77. Bloomfield, *Communication Strategies for Engaging Climate Skeptics*.

78. National Research Council, *Origin and Evolution of Earth: Research Questions for a Changing Planet* (Washington, DC: National Academies Press, 2008), 3, https://doi.org/10.17226/12161.

79. John Archibald, *One Plus One Equals One: Symbiosis and the Evolution of Complex Life* (Oxford: Oxford University Press, 2014), 6.

80. Stephen Meyer and Keith Fox, "Signature in the Cell—Stephen C. Meyer vs. Keith Fox," Unbelievable?, November 2011, audio, https://unbelievable.podbean.com/e/classic-replay-signature-in-the-cell-stephen-c-meyer-vs-keith-fox/; Randy Guliuzza and Karl Giberson, "Did God Use Evolution?: A Debate between Biblical

Creation and Theistic Evolution," Creation—Evolution Debate, September 30, 2013, http://www.socalsem.edu/divisions/center-for-creation-studies/creation-evolution-debate.

81. John Lyne and Henry F. Howe, "'Punctuated Equilibria': Rhetorical Dynamics of a Scientific Controversy," *Quarterly Journal of Speech* 72, no. 2 (1986): 132–47, https://doi.org/10.1080/00335638609383764.

82. Charles Darwin, *On the Origin of Species,* anniv. ed. (1859; repr.; New York: Signet, 2003), 53.

83. Smithsonian National Museum of Natural History, "Frequently Asked Questions," Smithsonian Institution's Human Origins Program, December 22, 2009, http://humanorigins.si.edu/education/frequently-asked-questions.

84. Countries that are part of the Commonwealth, primarily former British colonies, tend to use long-form numbers such as "thousand million." The example in this sentence is from an Australian publication, and the spiral map from the Jurassic Coast Trust (figure 5) is from the United Kingdom. Australian Academy of Science, "The Origins of Life on Earth," January 6, 2019, https://www.science.org.au/curious/space-time/origins-life-earth; "Names of Large Numbers," in *Simple English Wikipedia, the Free Encyclopedia,* accessed October 17, 2022, https://simple.wikipedia.org/w/index.php?title=Names_of_large_numbers&oldid=8494582.

85. The first four examples are from the National Academy of Sciences, *Science, Evolution, and Creationism* (Washington, DC: National Academies Press, 2008): 21, https://doi.org/10.17226/11876. The subsequent examples are from, in order, Bill Nye, *Undeniable: Evolution and the Science of Creation* (New York: St. Martin's Press, 2014); NASA, "NASA Research Reveals Major Insight into Evolution of Life on Earth," August 19, 2009, https://www.nasa.gov/topics/earth/features/astrobiology.html; Smithsonian National Museum of Natural History, "Frequently Asked Questions"; and National Academy of Sciences, *Science and Creationism.*

86. Hunt and Taylor, "Values and Cost–Benefit Analysis," 197–211; Pahl et al., "Perceptions of Time in Relation to Climate Change," 375–88; Tonn, Hemrick, and Conrad, "Cognitive Representations of the Future," 810–29.

87. Emma Frances Bloomfield, "Ark Encounter as Material Apocalyptic Rhetoric: Contemporary Creationist Strategies on Board Noah's Ark," *Southern Communication Journal* 82, no. 5 (2017): 263–77, https://doi.org/10.1080/1041794X.2017.1360384; Emma Frances Bloomfield, "Sensory Engagement with the Rhetoric of Science: Creationist *Copia* at the Discovery Center for Science and Earth History," *Rhetoric & Public Affairs* 25, no. 4 (2022): 65–93, https://muse.jhu.edu/article/885165.

88. "Bill Nye Debates Ken Ham," YouTube, February 4, 2014, informational video, 2:31:18, https://www.youtube.com/watch?v=z6kgvhG3AkI.

89. “Bill Nye Debates Ken Ham.”

90. Bloomfield, “Sensory Engagement with the Rhetoric of Science,” 65–93.

91. Daniel A. Salmon et al., “Vaccine Hesitancy: Causes, Consequences, and a Call to Action,” *Vaccine* 33 (2015): 67, https://doi.org/10.1016/j.vaccine.2015.09.035.

92. Lynda Walsh, “Before Climategate: Visual Strategies to Integrate Ethos across the ‘Is/Ought’ Divide in the IPCC’s Climate Change 2007: Summary for Policy Makers,” *Poroi* 6, no. 2 (2010): 39, https://doi.org/10.13008/2151-2957.1066.

93. *P*-values are measures of statistical significance that show the likelihood that a given outcome is due to chance. Depending on the field, p-values of 0.10 or 0.05 are deemed “significant,” which is interpreted as there is less than a 10 percent or 5 percent likelihood, respectively, that the differences observed in the data occurred due to random chance.

94. Lynda Walsh, *Scientists as Prophets: A Rhetorical Genealogy* (New York: Oxford University Press, 2013).

95. Walsh called the movement between scientific “is” statements and political “ought” statements an “upward pull toward action on all public scientific statements.” Walsh, *Scientists as Prophets,* 89.

96. G. Thomas Goodnight, “Public Discourse,” *Critical Studies in Media Communication* 4, no. 4 (1987): 428–32, https://doi.org/10.1080/15295038709360154; Walter R. Fisher, “Narrative Rationality and the Logic of Scientific Discourse,” *Argumentation* 8, no. 1 (1994): 21–32, https://doi.org/10.1007/BF00710701.

97. Latour outlined a similar boundary he called the science/politics divide. Walsh, “Before Climategate,” 33–61; Bruno Latour, “Telling Friends from Foes in the Time of the Anthropocene,” in *The Anthropocene and the Global Environmental Crisis,* ed. Clive Hamilton, François Gemenne, and Christophe Bonneuil (New York: Routledge, 2015), 145–55.

98. Cited in Pahl et al., “Perceptions of Time in Relation to Climate Change,” 376.

99. National Academy of Sciences and the Royal Society, “Climate Change.”

100. American Physical Society, “15.3 Statement on Earth’s Changing Climate,” November 14, 2015, http://www.aps.org/policy/statements/15_3.cfm.

101. NASA Global Climate Change, “The Causes of Climate Change,” accessed June 8, 2020, https://climate.nasa.gov/causes.

102. American Meteorological Society, “Climate Change: An Information Statement.”

103. M. Jimmie Killingsworth and Jacqueline S. Palmer, “The Discourse of ‘Environmentalist Hysteria,’” *Quarterly Journal of Speech* 81, no. 1 (1995): 8, https://doi.org/10.1080/00335639509384094; Brigitte Nerlich, “‘Climategate’: Paradoxical Metaphors and Political Paralysis,” *Environmental Values* 19, no. 4 (2010): 430, https://doi.org/10.3197/096327110X531543.

104. DiCaglio, *Scale Theory*, 187.

105. Emma Frances Bloomfield and Randall A. Lake, "Negotiating the End of the World in Climate Change Rhetoric: Climate Skepticism, Science, and Arguments," in *Communication for the Commons: Revisiting Participation and the Environment*, ed. Mark S. Meisner, Nadarajah Sriskandarajah, and Stephen P. Depoe (Turtle Island: International Environmental Communication Association, 2015).

106. H. Sterling Burnett, "Systematic Errors in Climate Models Make Their Predictions Worthless," Heartland Institute, November 26, 2019, https://soundcloud.com/user-694711047/systematic-errors-in-climate-models-make-their-predictions-worthless-guest-patrick-frank-phd?in=user-694711047/sets/environment-and-climate-news.

107. E. Calvin Beisner et al., "A Biblical Perspective on Environmental Stewardship," Acton Institute, accessed August 7, 2018, https://acton.org/public-policy/environmental-stewardship/theology-e/biblical-perspective-environmental-stewardship.

108. Ronald Stein, "Greta Preaches Many of the First Earth Day's Failed Predictions," Heartland Institute, March 23, 2020, https://www.heartland.org/news-opinion/news/greta-preaches-many-of-the-first-earth-days-failed-predictions.

109. Naomi Oreskes and Erik M. Conway, "Defeating the Merchants of Doubt," *Nature* 465 (2010): 686–87, https://www.nature.com/articles/465686a.

110. Centers for Disease Control, "Understanding Adverse Events and Side Effects," December 12, 2018, https://www.cdc.gov/vaccinesafety/ensuringsafety/sideeffects/index.html.

111. "Meet the Children," VaxTruth, August 9, 2011, http://vaxtruth.org/meet-the-children/.

112. "Meet the Children," VaxTruth. See also Lawrence on the "irreparable." Heidi Y. Lawrence, "Fear of the Irreparable: Narratives in Vaccination Rhetoric," *Narrative Inquiry in Bioethics* 6, no. 3 (2016): 205–9, https://doi.org/10.1353/nib.2016.0060.

113. "Meet the Children," VaxTruth.

114. John Tamny, *When Politicians Panicked: The New Coronavirus, Expert Opinion, and a Tragic Lapse of Reason* (Nashville, TN: Post Hill Press, 2021).

115. Centers for Disease Control, "COVID-19 and Your Health," last updated February 11, 2020, https://www.cdc.gov/coronavirus/2019-ncov/your-health/about-covid-19/basics-covid-19.html.

116. Ed Feil, "Can Scientists Predict All of the Ways the Coronavirus Will Evolve?," *The Conversation*, May 5, 2021, http://theconversation.com/can-scientists-predict-all-of-the-ways-the-coronavirus-will-evolve-156673.

117. Sam Levin, "How to Fight 'Covid Fatigue' as America Heads for a Deadly Winter," *The Guardian,* November 22, 2020, https://www.theguardian.com/world/2020/nov/22/coronavirus-how-to-convince-loved-ones-stay-home-masks.

118. Melissa Tate (@TheRightMelissa), "Looks like all they have all the variants planned out for the next 2 years SMH," Twitter, July 29, 2021, 11:35 a.m., https://archive.ph/DadJr.

119. Reuters Fact Check, "Fact Check-Bogus Document Claims to Show 'Release Dates' of COVID-19 Variants," *Reuters,* August 2, 2021, https://www.reuters.com/article/factcheck-variants-chart-idUSL1N2P91AX.

120. Adam Cancryn and Krista Mahr, "Biden Declared the Pandemic 'Over.' His Covid Team Says It's More Complicated," *POLITICO,* September 19, 2022, https://www.politico.com/news/2022/09/19/biden-pandemic-over-covid-team-response-00057649.

121. Cancryn and Mahr; Centers for Disease Control, "COVID Data Tracker Weekly Review," last updated September 16, 2022, https://www.cdc.gov/coronavirus/2019-ncov/covid-data/covidview/index.html.

122. Katie Bach, "New Data Shows Long Covid Is Keeping as Many as 4 Million People Out of Work," Brookings Institution, August 24, 2022, https://www.brookings.edu/research/new-data-shows-long-covid-is-keeping-as-many-as-4-million-people-out-of-work/.

123. Pezzullo, "Performing Critical Interruptions," 1–25.

CHAPTER 4

1. Scott Waldman and Nina Heikkinen, "As Climate Scientists Speak Out, Sexist Attacks Are on the Rise," *Scientific American,* August 22, 2018, https://www.scientificamerican.com/article/as-climate-scientists-speak-out-sexist-attacks-are-on-the-rise/.

2. Global Witness, "Global Hating: How Online Abuse of Climate Scientists Harms Climate Action," April 4, 2023, https://www.globalwitness.org/en/campaigns/digital-threats/global-hating/.

3. Virginia Gewin, "Real-Life Stories of Online Harassment—and How Scientists Got through It," *Nature* 562, no. 7727 (2018): 449–50, https://doi.org/10.1038/d41586-018-07046-0.

4. James P. Zappen, "Digital Rhetoric: Toward an Integrated Theory," *Technical Communication Quarterly* 14, no. 3 (2005): 321, https://doi.org/10.1207/s15427625tcq1403_10.

5. Emma Frances Bloomfield and Denise Tillery, "The Circulation of Climate Change Denial Online: Rhetorical and Networking Strategies on Facebook,"

Environmental Communication 13, no. 1 (2019): 23–34, https://doi.org/10.1080/17524032.2018.1527378.

6. Editorial, "COVID Scientists in the Public Eye Need Protection from Threats," *Nature* 598, no. 7880 (2021): 236, https://doi.org/10.1038/d41586-021-02757-3.

7. Tricia R. Pendergrast et al., "Prevalence of Personal Attacks and Sexual Harassment of Physicians on Social Media," *JAMA Internal Medicine* 181, no. 4 (2021): 550–52, https://doi.org/10.1001/jamainternmed.2020.7235.

8. The experiences of Kuppalli, Van Ranst, and Whitty are detailed by Nogrady. Bianca Nogrady, "'I Hope You Die': How the COVID Pandemic Unleashed Attacks on Scientists," *Nature* 598, no. 7880 (2021): 250–53, https://doi.org/10.1038/d41586-021-02741-x.

9. Sarah Johnson et al., "'They Stormed the ICU and Beat the Doctor': Health Workers under Attack," *The Guardian*, June 7, 2021, https://www.theguardian.com/global-development/2021/jun/07/they-stormed-the-icu-and-beat-the-doctor-health-workers-under-attack.

10. See Graham Dixon, Jay Hmielowski, and Yanni Ma, "Improving Climate Change Acceptance among U.S. Conservatives through Value-Based Message Targeting," *Science Communication* 39, no. 4 (2017): 520–34, https://doi.org/10.1177/1075547017715473; Melissa L. Carrion, "'You Need to Do Your Research': Vaccines, Contestable Science, and Maternal Epistemology," *Public Understanding of Science* 27, no. 3 (2018): 310–24, https://doi.org/10.1177/0963662517728024; Brian McCammack, "Hot Damned America: Evangelicalism and the Climate Change Policy Debate," *American Quarterly* 59, no. 3 (2007): 645–68, https://doi.org/10.1353/aq.2007.0065.

11. Denise Tillery and Emma Frances Bloomfield, "Hyperrationality and Rhetorical Constellations in Digital Climate Change Denial: A Multi-Methodological Analysis of the Discourse of Watts Up with That," *Technical Communication Quarterly* 31, no. 4 (2022): 356–73, https://doi.org/10.1080/10572252.2021.2019317.

12. Margaret Kovach, *Indigenous Methodologies: Characteristics, Conversations, and Contexts*, repr. ed. (Toronto: University of Toronto Press, 2010), 94.

13. Kovach, 94; see also James Phelan, *Narrative as Rhetoric: Technique, Audiences, Ethics, Ideology* (Columbus: Ohio State University Press, 1996).

14. McClure proposed the term *narrative identification* to make sense of why people may adopt stories in the context of creationism despite their having low probability and fidelity compared to the stories of evolution that are supported by material evidence. Kevin McClure, "Resurrecting the Narrative Paradigm: Identification and the Case of Young Earth Creationism," *Rhetoric Society Quarterly* 39, no. 2 (2009): 189–211, https://doi.org/10.1080/02773940902766771.

15. Bonnie J. Dow, "Taking Trump Seriously: Persona and Presidential Politics in 2016," *Women's Studies in Communication* 40, no. 2 (2017): 136, https://doi.org/10.1080/07491409.2017.1302258.

16. Davey Alba, "Virus Conspiracists Elevate a New Champion," *New York Times*, May 9, 2020, https://www.nytimes.com/2020/05/09/technology/plandemic-judy-mikovitz-coronavirus-disinformation.html; Matthew Brown, "Fact Check: 'Plandemic' Sequel Makes False Claims about Bill Gates," *USA Today*, August 25, 2020, https://www.usatoday.com/story/news/factcheck/2020/08/25/fact-check-plandemic-sequel-makes-false-claims-bill-gates/5627223002/; Kevin Stankiewicz, "Bill Gates: Vaccine Conspiracies Targeting Dr. Fauci and Me Are 'Unfortunate' and Hurt Public Trust," *CNBC*, October 14, 2020, https://www.cnbc.com/2020/10/14/bill-gates-anti-vaxxer-theories-about-fauci-and-me-hurt-public-trust.html.

17. Walter R. Fisher, "Narrative Rationality and the Logic of Scientific Discourse," *Argumentation* 8, no. 1 (1994): 24, https://doi.org/10.1007/BF00710701.

18. Fisher, 24.

19. Fisher, 24.

20. Pamela Pietrucci and Leah Ceccarelli, "Scientist Citizens: Rhetoric and Responsibility in L'Aquila," *Rhetoric and Public Affairs* 22, no. 1 (2019): 95–128, https://muse.jhu.edu/article/718702.

21. Doug Cloud, "The Corrupted Scientist Archetype and Its Implications for Climate Change Communication and Public Perceptions of Science," *Environmental Communication* 14, no. 6 (2020): 816–29, https://doi.org/10.1080/17524032.2020.1741420.

22. Robert K. Merton, *The Sociology of Science: Theoretical and Empirical Investigations*, ed. Norman W. Storer (Chicago: University of Chicago Press, 1979).

23. As discussed in chapter 2, disinterestedness is a feature of the scientific enterprise itself, not of individual scientists. Therefore, pointing out the financial or political motivations of individual storytellers would not necessarily violate the norm as Robert Merton defined it. However, many people see disinterestedness operating at an individual level, which creates expectations of scientists as martyrs committed to a vow of poverty and solely driven by scientific inquiry. Richard Wunderlich, "The Scientific Ethos: A Clarification," *British Journal of Sociology* 25, no. 3 (1974): 373–77, https://doi.org/10.2307/589402.

24. As a personal example, I have some academic friends in the sciences who do not vote because they see political neutrality as a responsibility of scientists. Bruno Latour, "Telling Friends from Foes in the Time of the Anthropocene," in *The Anthropocene and the Global Environmental Crisis*, ed. Clive Hamilton, François Gemenne, and Christophe Bonneuil (New York: Routledge, 2015), 145–55.

25. Fisher, "Narrative Rationality and the Logic of Scientific Discourse," 24.

26. Fisher, 24.

27. Fisher, 24.

28. Kenneth Burke, *Permanence and Change: An Anatomy of Purpose* (Berkeley: University of California Press, 1935), 37.

29. Amy Goldstein and Emily Guskin, "Almost One-Third of Black Americans Know Someone Who Died of Covid-19, Survey Shows," *Washington Post,* June 26, 2020, https://www.washingtonpost.com/health/almost-one-third-of-black-americans-know-someone-who-died-of-covid-19-survey-shows/2020/06/25/3ec1d4b2-b563-11ea-aca5-ebb63d27e1ff_story.html.

30. Jeanne Fahnestock, "Accommodating Science: The Rhetorical Life of Scientific Facts," *Written Communication* 15, no. 3 (1998): 330–50, https://doi.org/10.1177/0741088386003003001.

31. Fahnestock, 344.

32. Ceccarelli uses the phrase "mainstream scientists and their allies," which I group together as science storytellers. Leah Ceccarelli, "Manufactured Scientific Controversy: Science, Rhetoric, and Public Debate," *Rhetoric & Public Affairs* 14, no. 2 (2011): 198, https://doi.org/10.1353/rap.2010.0222.

33. In my first book, I spoke to climate skeptics who were having difficulty navigating the perceived sea of experts speaking out about climate change. In a conversation I had with Joseph (pseudonym), he stated that he was confused about "which PhD is the right one" when doing research and finding conflicting information. In a conversation with Victor (pseudonym), he argued that climate change must not be truly correct because if it were, there would not be a "single defector" to the science. Emma Frances Bloomfield, *Communication Strategies for Engaging Climate Skeptics: Religion and the Environment* (New York: Routledge, 2019), 85, 101.

34. Ceccarelli, "Manufactured Scientific Controversy," 198.

35. E. Calvin Beisner, "Three Great Legacies of Climategate," *Cornwall Alliance,* December 3, 2014, https://cornwallalliance.org/2014/12/three-great-legacies-of-climategate/.

36. Beisner.

37. Joseph L. Bast, "The Myth of the 98 Percent," Heartland Institute, October 2012, https://www.heartland.org/_template-assets/documents/publications/10-2012_myth_of_the_98_percent.pdf.

38. Jerry Bergman, "Contemporary Suppression of the Theistic Worldview," Answers in Genesis, August 1, 1995, https://answersingenesis.org/intelligent-design/contemporary-suppression-of-the-theistic-worldview/.

39. Robert T. Pennock, "Creationism and Intelligent Design," *Annual Review of Genomics and Human Genetics* 4, no. 1 (2003): 156, https://doi.org/10.1146/annurev.genom.4.070802.110400.

40. Institute for Creation Research, "ICR's Approach to Scientific Investigation," accessed October 8, 2022, https://www.icr.org/how-we-do-research.

41. Christine Dao, "Evolution Controversy's Outdated, One-Sided Exposure," Institute for Creation Research, July 26, 2011, https://www.icr.org/article/evolution-controversys-outdated-one.

42. Henry M. Morris, "Evolution Is Religion—Not Science," Institute for Creation Research, February 1, 2001, https://www.icr.org/article/455/.

43. American Academy of Pediatrics, *Vaccine Advocacy Snapshot,* 2020, https://downloads.aap.org/DOFA/AAPVaccinationsOnePager.pdf.

44. Jeanne P. Spencer, Ruth H. Trondsen Pawlowski, and Stephanie Thomas, "Vaccine Adverse Events: Separating Myth from Reality," *American Family Physician* 95, no. 12 (2017): 786, https://www.aafp.org/pubs/afp/issues/2017/0615/p786.html.

45. Spencer, Trondsen Pawlowski, and Thomas, 786.

46. Missy Fluegge, "8 Reasons Why Your Child's Doctor Pushes Vaccines," VacTruth, August 6, 2016, https://vactruth.com/2016/08/06/8-reasons-doctors-push-vaccines/.

47. Fluegge.

48. Fluegge.

49. Fluegge.

50. Sarah Carrasco, "Vaccines: The Exception to Medical Coercion," VacTruth, July 27, 2016, https://vactruth.com/2016/07/27/vaccines-and-coercion/.

51. Carrasco.

52. Marcella Piper-Terry, "Vaccines, Autism, and Susceptible Groups: Is Your Child at Increased Risk?," VaxTruth, September 10, 2014, http://vaxtruth.org/2014/09/susceptible-groups/.

53. Marcella Piper-Terry, "Vaccines Are Poisoning African-American and Latino Babies—and the Rest of Our Children," VaxTruth, October 13, 2015, http://vaxtruth.org/2015/10/toxic-vaccines/.

54. Jeffrey John Aufderheide, "Who Says There Is No Money in Making Vaccines? At Least $5.7 Billion Given to Vaccine Manufacturers in 2011 by US Gov't," VacTruth, November 30, 2012, https://vactruth.com/2012/11/30/2011-billions-vaccine-manufacturers/.

55. Assertions such as these are based on the National Vaccine Injury Compensation Program (NVICP), which was created in part because of the increase in frivolous vaccine injury lawsuits in the 1980s that were costing manufacturers time and money that detracted from vaccine development and production and subsequently put national health at risk. As an alternative to adjudicating through traditional legal venues, the NVICP works through a process of petition-filing where individual cases are reviewed and compensation may be provided even if someone is not found to have been injured by a vaccine. This does not stop vaccine skeptical groups from touting NVICP payouts as evidence of legal culpability. U.S. Health Resources & Services Administration, "National Vaccine Injury Compensation Program," May 11, 2017, https://www.hrsa.gov/vaccine-compensation/index.html; Fluegge, "8 Reasons Why Your Child's Doctor Pushes Vaccines."

56. Carly Mayberry, "Hospital Workers Speak Out about COVID Protocols from Coast to Coast," *Epoch Times,* October 4, 2022, https://www.theepochtimes.com/health/hospital-workers-speak-out-about-covid-protocols-from-coast-to-coast_4773779.html.

57. The phrase "fell on deaf ears" is ableist because it links a lack of hearing capability with ignorance. A better phrase would be that the doctors were not listened to or their concerns were not taken into consideration. I have retained the language because it is within a direct quotation about how this particular doctor expressed their concerns. Mayberry, "Hospital Workers Speak Out."

58. Mayberry.

59. Tammy Bruce, "Flip-Flop Fauci: The Purveyor of Doom and Gloom," *Washington Times,* October 6, 2021, https://www.washingtontimes.com/news/2021/oct/6/flip-flop-fauci-the-purveyor-of-doom-and-gloom/; Opinion, "Did Fauci Flip-Flop or Follow the Science?," *Hickory Daily Record,* July 22, 2021, https://hickoryrecord.com/opinion/columnists/column-did-fauci-flip-flop-or-follow-the-science/article_eb57e816-ea40–11eb-8032–43e9e70a535b.html; Brittany Bernstein, "Fauci Flip Flops on Lab-Leak Theory: 'Highly Likely' COVID-19 Originated Naturally," *National Review,* May 25, 2021, https://www.nationalreview.com/news/fauci-flip-flops-on-lab-leak-theory-highly-likely-covid-19-originated-naturally/.

60. Tillery and Bloomfield, "Hyperrationality and Rhetorical Constellations," 357, 358.

61. Cloud, "The Corrupted Scientist Archetype," 819.

62. Denise Tillery, *Commonplaces of Scientific Evidence in Environmental Discourses* (New York: Routledge, 2018), 154.

63. Ileana Johnson Paugh, "The Global Warming/Climate Change Alarmists: Forward or Bust," *Canada Free Press,* February 10, 2013, http://www.canadafreepress.com/index.php/article/the-global-warming-climate-change-alarmists-forward-or-bust.

64. Michael Shellenberger, "On Behalf of Environmentalists, I Apologize for the Climate Scare," *Environmental Progress,* June 29, 2020, https://environmentalprogress.org/big-news/2020/6/29/on-behalf-of-environmentalists-i-apologize-for-the-climate-scare.

65. Stephen Moore, "Follow the (Climate Change) Money," Heritage Foundation, December 18, 2018, https://www.heritage.org/environment/commentary/follow-the-climate-change-money.

66. Moore.

67. Robert Rapier, "The Big Money of Environmentalism," *Christian Science Monitor,* July 28, 2012, https://www.csmonitor.com/Business/Consumer-Energy-Report/2012/0728/The-big-money-of-environmentalism.

68. Bast, "The Myth of the 98 Percent."

69. James Murphy, "The New American: Al Gore Continues to Profit Off of the Climate Alarmism," Energy & Environment Legal Institute, May 20, 2022, https://eelegal.org/the-new-american/.

70. "Al Gore's Climate Change Hypocrisy Is as Big as His Energy-Sucking Mansion," *Investor's Business Daily,* August 3, 2017, https://www.investors.com/politics/editorials/al-gores-climate-change-hypocrisy-is-as-big-as-his-energy-sucking-mansion/.

71. Moore, "Follow the (Climate Change) Money."

72. Spencer Pond, "How Can It Be about the Money? Immunizations Are Free! Right?," VaxTruth, September 14, 2011, http://vaxtruth.org/2011/09/how-can-it-be-about-the-money-immunizations-are-free-right/.

73. Marcella Piper-Terry, "Dear Pregnant Mom Who Is Just Beginning to Question Vaccines," VaxTruth, May 14, 2016, http://vaxtruth.org/2016/05/dear-pregnant-mom/.

74. Aufderheide, "Who Says There Is No Money in Making Vaccines?"

75. Daniel Funke, "Facebook Posts Falsely Claim Dr. Fauci Has Millions Invested in a Coronavirus Vaccine," *PolitiFact,* April 15, 2020, https://www.politifact.com/factchecks/2020/apr/15/facebook-posts/facebook-posts-falsely-claim-dr-fauci-has-millions/.

76. *Plandemic* features an interview with Judy Mikovits, who calls Dr. Anthony Fauci "Tony" throughout the video. This could be a strategy to undermine Fauci's credibility through using a diminutive name or to establish that she and Fauci are close enough that she would be able to use an otherwise unused nickname. Martin Enserink and Jon Cohen, "Fact-Checking Judy Mikovits, the Controversial Virologist Attacking Anthony Fauci in a Viral Conspiracy Video," *Science,* May 8, 2020, https://www.sciencemag.org/news/2020/05/fact-checking-judy-mikovits-controversial-virologist-attacking-anthony-fauci-viral.

77. The repetition of false information, even when corrected, can still shape people's attitudes and behaviors on a topic. Thorson referred to the stickiness of false information as "belief echoes." Katherine Fung, "Rand Paul Spars with Fauci Over Money Tied to Vaccines," *Newsweek,* June 16, 2022, https://www.newsweek.com/rand-paul-fauci-spar-over-vaccine-money-1716661; Emily Thorson, "Belief Echoes: The Persistent Effects of Corrected Misinformation," *Political Communication* 33, no. 3 (2016): 460–80, https://doi.org/10.1080/10584609.2015.1102187.

78. Lee Fang, "Biden's Inner Circle Maintains Close Ties to Vaccine Makers, Disclosures Reveal," *The Intercept,* March 24, 2021, https://theintercept.com/2021/03/24/covid-vaccine-stocks-biden-conflict/.

79. Fang.

80. Fang.

81. Mayberry, "Hospital Workers Speak Out."

82. Mayberry.

83. Michelle Rogers, "Fact Check: Medicare Pays Hospitals More Money for COVID-19 Patients," *USA Today,* April 24, 2020, https://www.usatoday.com/story/news/factcheck/2020/04/24/fact-check-medicare-hospitals-paid-more-covid-19-patients-coronavirus/3000638001/.

84. Rogers.

85. Angelo Fichera, "Hospital Payments and the COVID-19 Death Count," *FactCheck.Org* (blog), April 21, 2020, https://www.factcheck.org/2020/04/hospital-payments-and-the-covid-19-death-count/.

86. Cloud, "The Corrupted Scientist Archetype," 821.

87. Tillery, *Commonplaces of Scientific Evidence*, 19.

88. Jordan J. Ballor, "Christian Reformed Church Backs UN Climate Change Agenda," Acton Institute, July 8, 2015, https://acton.org/pub/commentary/2015/07/08/christian-reformed-church-backs-un-climate-change-agenda; Kishore Jayabalan, "The Economic Reeducation of Pope Francis?," Acton Institute, September 30, 2015, https://www.acton.org/pub/commentary/2015/09/30/economic-reeducation-pope-francis.

89. Shellenberger, "On Behalf of Environmentalists."

90. Nongovernmental International Panel on Climate Change, "About the IPCC," accessed June 15, 2020, http://climatechangereconsidered.org/about-the-ipcc/.

91. E. Calvin Beisner, "World Magazine Exposes Evangelical Environmentalists' Growing Dependence on Green-Left Funding," Cornwall Alliance, May 30, 2015,https://cornwallalliance.org/2015/05/world-magazine-exposes-evangelical-environmentalists-growing-dependence-on-green-left-funding/.

92. Heartland Institute, "Action Plan for President Trump," November 21, 2016, https://www.heartland.org/topics/government-politics/trump-action-plan/; Paul Driessen, "The Green New Deal Dress Rehearsal," Heartland Institute, June 8, 2020, https://www.heartland.org/news-opinion/news/the-green-new-deal-dress-rehearsal.

93. Driessen.

94. Bloomfield and Tillery, "The Circulation of Climate Change Denial Online."

95. Driessen, "The Green New Deal Dress Rehearsal."

96. Driessen.

97. Cornwall Alliance, "An Evangelical Declaration on Global Warming," *For the Stewardship of Creation* (blog), May 1, 2009, http://www.cornwallalliance.org/articles/read/an-evangelical-declaration-on-global-warming/.

98. Roger Patterson, "What Is Science?," Answers in Genesis, February 22, 2007, https://answersingenesis.org/what-is-science/what-is-science/.

99. Patricia Engler, "Study Finds Liberal Bias at Canadian, American, and British Universities," Answers in Genesis, March 24, 2021, https://answers

ingenesis.org/blogs/patricia-engler/2021/03/24/study-finds-liberal-bias-canadian-american-british-universities/.

100. Ken Ham, "Are Christian Ministries 'Hate Groups' for Believing the Bible?," Answers in Genesis, May 26, 2017, https://answersingenesis.org/blogs/ken-ham/2017/05/26/are-christian-ministries-hate-groups-for-believing-bible/.

101. Ken Ham and Bodie Hodge, "Atheism Is Religion," Answers in Genesis, February 20, 2017, https://answersingenesis.org/world-religions/atheism/atheism-is-religion/; see also Morris, "Evolution Is Religion—Not Science."

102. Ken Ham, "Smithsonian Exhibition—Deception and Atheism," Answers in Genesis, July 26, 2010, https://answersingenesis.org/human-evolution/smithsonian-exhibition-deception-and-atheism/.

103. Ham.

104. John Altevogt, "Guest Column: 'The Theory of Evolution,'" Answers in Genesis, January 17, 2000, https://answersingenesis.org/theory-of-evolution/the-theory-of-evolution/.

105. Altevogt.

106. Carroll Doherty, Seth Motel, and Rachel Weisel, "83% Say Measles Vaccine Is Safe for Healthy Children," Pew Research Center, February 9, 2015, https://www.pewresearch.org/politics/2015/02/09/83-percent-say-measles-vaccine-is-safe-for-healthy-children/.

107. Charles McCoy, "Anti-Vaccination Beliefs Don't Follow the Usual Political Polarization," *The Conversation,* August 23, 2017, https://theconversation.com/anti-vaccination-beliefs-dont-follow-the-usual-political-polarization-81001; see also Doherty, Motel, and Weisel, "83% Say Measles Vaccine Is Safe."

108. P. Sol Hart, Sedona Chinn, and Stuart Soroka, "Politicization and Polarization in COVID-19 News Coverage," *Science Communication* 42, no. 5 (2020): 679–97, https://doi.org/10.1177/1075547020950735.

109. Lauren Egan, "Trump Calls Coronavirus Democrats' 'New Hoax,'" *NBC News,* February 28, 2020, https://www.nbcnews.com/politics/donald-trump/trump-calls-coronavirus-democrats-new-hoax-n1145721.

110. Nicholas S. Paliewicz and Emma Frances Bloomfield, "Of Markets, Masks, and (White) Men: Mimetic Performances of Parasitic Publicity during the COVID-19 Pandemic," *Women's Studies in Communication* 46, no. 1 (2023): 66, https://doi.org/10.1080/07491409.2022.2156416.

111. Dareh Gregorian, "Marjorie Taylor Greene Hit with Four More Fines for Refusing to Wear a Mask," *NBC News,* November 1, 2021, https://www.nbcnews.com/politics/congress/marjorie-taylor-greene-hit-four-more-fines-refusing-wear-mask-n1282902.

112. Tom Hals, "Republican Governors Lead Attack on Biden Vaccine Mandate," *Reuters,* November 5, 2021, https://www.reuters.com/world/us/republican-governors-lead-attack-biden-vaccine-mandate-2021-11-05/; Myah Ward,

"Coronavirus Lockdowns Pit Republican Governors against Democratic Mayors," *Politico,* April 21, 2020, https://www.politico.com/news/2020/04/21/coronavirus-pits-republican-governors-democratic-mayors-198858.

113. Ward, "Coronavirus Lockdowns."

114. Katie Rogers, Lara Jakes, and Ana Swanson, "Trump Defends Using 'Chinese Virus' Label, Ignoring Growing Criticism," *New York Times,* March 18, 2020, https://www.nytimes.com/2020/03/18/us/politics/china-virus.html; "President Trump Calls Coronavirus 'Kung Flu,'" *BBC News,* June 24, 2020, https://www.bbc.com/news/av/world-us-canada-53173436; Robert Hart, "Trump's 'Chinese Virus' Tweet Helped Fuel Anti-Asian Hate on Twitter, Study Finds," *Forbes,* March 19, 2021, https://www.forbes.com/sites/roberthart/2021/03/19/trumps-chinese-virus-tweet-helped-fuel-anti-asian-hate-on-twitter-study-finds/.

115. Zachery Eanes, "Chinese State Media Targets UNC Lab as US Investigates Origin of COVID-19 Pandemic," *Raleigh News & Observer,* August 26, 2021, https://www.newsobserver.com/news/coronavirus/article253740393.html.

116. Eanes; Tom Kertscher and Noah Y. Kim, "Fact Checking Rand Paul's Claims about UNC Researcher, COVID-19 Origins," *WRAL,* May 18, 2021, https://www.wral.com/fact-checking-rand-paul-s-claims-about-unc-researcher-covid-19-origins/19684174/.

117. Peter Rudiak-Gould, "'We Have Seen It with Our Own Eyes': Why We Disagree about Climate Change Visibility," *Weather, Climate, and Society* 5, no. 2 (2013): 120–32, https://doi.org/10.1175/WCAS-D-12-00034.1.

118. Dylan M. Harris, "Telling Stories about Climate Change," *Professional Geographer* 72, no. 3 (2020): 314, https://doi.org/10.1080/00330124.2019.1686996.

119. Harris, 314.

120. "Cornwall Alliance," accessed April 6, 2018, https://cornwallalliance.org.

121. David C. Barker and David H. Bearce, "End-Times Theology, the Shadow of the Future, and Public Resistance to Addressing Global Climate Change," *Political Research Quarterly* 66, no. 2 (2013): 267–79, https://doi.org/10.1177/1065912912442243.

122. Cornwall Alliance, "The Biblical Perspective of Environmental Stewardship: Subduing and Ruling the Earth to the Glory of God and the Benefit of Our Neighbors," accessed April 2, 2018, https://cornwallalliance.org/landmark-documents/the-biblical-perspective-of-environmental-stewardship-subduing-and-ruling-the-earth-to-the-glory-of-god-and-the-benefit-of-our-neighbors/.

123. E. Calvin Beisner, "Putting Together the Pieces in the Spiritual World War," Cornwall Alliance, October 1, 2012, https://cornwallalliance.org/2012/10/putting-together-the-pieces-in-the-spiritual-world-war/.

124. Ballor, "Christian Reformed Church Backs UN Climate Change Agenda."

125. E. Calvin Beisner et al., "A Biblical Perspective on Environmental Stewardship," Acton Institute, accessed August 7, 2018, https://acton.org/public-policy/environmental-stewardship/theology-e/biblical-perspective-environmental-stewardship.

126. Joseph Bast, "Pro-Environment, Pro-Energy, and Pro-Jobs," Heartland Institute, August 10, 2016, https://www.heartland.org/publications-resources/publications/policy-tip-sheet-pro-environment-pro-energy-and-pro-jobs.

127. William J. Kinsella et al., "Narratives, Rhetorical Genres, and Environmental Conflict: Responses to Schwarze's 'Environmental Melodrama,'" *Environmental Communication* 2, no. 1 (2008): 95, https://doi.org/10.1080/17524030801980242.

128. "Bill Nye Debates Ken Ham," YouTube, February 4, 2014, informational video, 2:31:18, https://www.youtube.com/watch?v=z6kgvhG3AkI.

129. Ken Ham, "How to Prevent Losing Another Generation," Answers in Genesis, September 6, 2015, https://answersingenesis.org/christianity/church/how-prevent-losing-another-generation/.

130. Ken Ham, "Christianity: Child Abuse? A Virus?" Answers in Genesis, March 1, 2006, https://answersingenesis.org/train-up-a-child/christian-parenting/christianity-child-abuse-a-virus/.

131. National Center for Science Education, *Voices for Evolution,* 3rd ed. (2008), vi, https://ncse.ngo/voices-evolution-0.

132. Donald R. Prothero, *Evolution: What the Fossils Say and Why It Matters,* 2nd ed. (New York: Columbia University Press, 2017): xi, xvii.

133. "Bill Nye Debates Ken Ham."

134. Prothero, *Evolution,* xii.

135. Geoffrey M. Hodgson, "Social Darwinism in Anglophone Academic Journals: A Contribution to the History of the Term," *Journal of Historical Sociology* 17, no. 4 (2004): 428–63, https://doi.org/10.1111/j.1467-6443.2004.00239.x; Diane B. Paul, "Darwin, Social Darwinism and Eugenics," in *The Cambridge Companion to Darwin,* ed. Jonathan Hodge and Gregory Radick (Cambridge: Cambridge University Press, 2003), 214–38; Emma Frances Bloomfield, "The Rhetoric of Energy Darwinism: Neoliberal Piety and Market Autonomy in Economic Discourse," *Rhetoric Society Quarterly* 49, no. 4 (2019): 320–41, https://doi.org/10.1080/02773945.2019.1634831.

136. Pew Research Center, "Overview: The Conflict between Religion and Evolution," February 3, 2014, https://www.pewforum.org/2009/02/04/overview-the-conflict-between-religion-and-evolution/.

137. Bill Nye, *Undeniable: Evolution and the Science of Creation* (New York: St. Martin's Press, 2014), 19.

138. Understanding Evolution, *Misconceptions about Evolution* (2012), https://evolution.berkeley.edu/evolibrary/misconceptions_about_evolution.pdf.

139. Pew Research Center, "Overview."

140. National Research Council, *Thinking Evolutionarily: Evolution Education across the Life Sciences—Summary of a Convocation* (Washington, DC: National Academies Press, 2012), 39, 40.

141. Kristy Maddux, "Fundamentalist Fool or Populist Paragon? William Jennings Bryan and the Campaign against Evolutionary Theory," *Rhetoric and Public Affairs* 16, no. 3 (2013): 502, https://doi.org/10.14321/rhetpublaffa.16.3.0489.

142. Jordynn Jack, *Autism and Gender: From Refrigerator Mothers to Computer Geeks* (Champaign: University of Illinois Press, 2014); Carrion, "'You Need to Do Your Research'," 310–24; Stephanie Willes, "The Affordances of the Internet and the Creation of Villains and Heroes in Anti-Vaccine Stories: A Narrative Criticism of VaxTruth" (master's thesis, University of Nevada, Las Vegas, 2019).

143. Piper-Terry, "Dear Pregnant Mom."

144. Piper-Terry.

145. In this meta-analysis, the authors compiled data from studies assessing the relationship between vaccines and autism that, in total, involved more than 1.2 million children. The meta-analysis conclusively found "no relationship between vaccination and autism," and "no evidence for increased risk of developing autism or ASD [an autism spectrum disorder] following MMR [measles, mumps, and rubella vaccination]." Luke E. Taylor, Amy L. Swerdfeger, and Guy D. Eslick, "Vaccines Are Not Associated with Autism: An Evidence-Based Meta-Analysis of Case-Control and Cohort Studies," *Vaccine* 32, no. 29 (2014): 3623–29, https://doi.org/10.1016/j.vaccine.2014.04.085.

146. Shari Hoppin, "Applying the Narrative Paradigm to the Vaccine Debates," *American Communication Journal* 18, no. 2 (2016): 45–55.

147. Hoppin, 52.

148. Trip Gabriel, "'Magic' Weight-Loss Pills and Covid Cures: Dr. Oz under the Microscope," *New York Times,* December 26, 2021, https://www.nytimes.com/2021/12/26/us/politics/dr-oz-medical-advice.html.

149. Gabriel.

150. Jonathan Jarry, "Science vs. Joe Rogan," McGill Office for Science and Society, November 20, 2021, https://www.mcgill.ca/oss/article/covid-19-health-and-nutrition-pseudoscience/science-vs-joe-rogan.

151. Jarry.

152. Jarry.

153. David Crow and Patti Waldmeir, "US Anti-Lockdown Protests: 'If You Are Paranoid about Getting Sick, Just Don't Go Out,'" *Financial Times,* April 22, 2020, https://www.ft.com/content/15ca3a5f-bc5c-44a3-99a8-c446f6f6881c.

154. Paul Ratner, "Liberty or Death? The Coronavirus Attacks the Soul of America," *Big Think,* April 15, 2020, https://bigthink.com/the-present/liberty-or-death-the-coronavirus-attacks-the-soul-of-america/.

155. Crow and Waldmeir, "US Anti-Lockdown Protests."

156. Bess Levin, "Texas Lt. Governor: Old People Should Volunteer to Die to Save the Economy," *Vanity Fair*, March 24, 2020, https://www.vanityfair.com/news/2020/03/dan-patrick-coronavirus-grandparents.

CHAPTER 5

1. Catie Keck, "Don't Look Up Narrowly Misses Becoming Netflix's All-Time-Best Film Debut," *The Verge*, January 25, 2022, https://www.theverge.com/2022/1/25/22878712/netflix-dont-look-up-red-notice-bird-box-top-10.

2. *Don't Look Up*, Rotten Tomatoes (reviews), accessed October 13, 2022, https://www.rottentomatoes.com/m/dont_look_up_2021.

3. "*Don't Look Up*," Wikipedia, accessed October 13, 2022, https://en.wikipedia.org/wiki/Don%27t_Look_Up.

4. Dylan Connor, "Why *Don't Look Up* Has Aggressively Divided Critics & Audiences," *ScreenRant*, January 1, 2022, https://screenrant.com/dont-look-up-critics-audiences-reactions-divisive-reason/.

5. Julie Doyle, "Communicating Climate Change in 'Don't Look Up,'" *Journal of Science Communication* 21, no. 5 (2022): 2, https://doi.org/10.22323/2.21050302.

6. Doyle, 2.

7. Peter Friederici, *Beyond Climate Breakdown: Envisioning New Stories of Radical Hope* (Cambridge, MA: MIT Press, 2022).

8. Michael D. Jones and Holly Peterson, "Narrative Persuasion and Storytelling as Climate Communication Strategies," *Oxford Research Encyclopedia of Climate Science* (2017), https://doi.org/10.1093/acrefore/9780190228620.013.384.

9. Angeline Sangalang and Emma Frances Bloomfield, "Mother Goose and Mother Nature: Designing Stories to Communicate Information about Climate Change," *Communication Studies* 69, no. 5 (2018): 585, https://doi.org/10.1080/10510974.2018.1489872.

10. Steven Schwarze, "Environmental Melodrama," *Quarterly Journal of Speech* 92, no. 3 (2006): 239–61, https://doi.org/10.1080/00335630600938609.

11. Leah Ceccarelli, "CRISPR as Agent: A Metaphor That Rhetorically Inhibits the Prospects for Responsible Research," *Life Sciences, Society and Policy* 14 (2018): 10, https://doi.org/10.1186/s40504-018-0088-8.

12. Ceccarelli, 1–13.

13. As discussed in chapter 1, the term *scientist citizen* is a conceptualization of the role of scientists as existing in the technical and public spheres and thus operating under both scientific and public commitments. See Pamela Pietrucci and Leah Ceccarelli, "Scientist Citizens: Rhetoric and Responsibility in L'Aquila," *Rhetoric and Public Affairs* 22, no. 1 (2019): 95–128, https://muse.jhu.edu/article/718702.

14. Jean Goodwin and Michael F. Dahlstrom, "Communication Strategies for Earning Trust in Climate Change Debates," *Wiley Interdisciplinary Reviews: Climate Change* 5, no. 1 (2014): 151, https://doi.org/10.1002/wcc.262.

15. Goodwin and Dahlstrom, 152.

16. Elisha M. Wood-Charlson et al., "Translating Science into Stories," *Limnology and Oceanography Bulletin* 24, no. 3 (2015): 73, emphasis in original.

17. Brian E. Martin and Wytze Brouwer, "The Sharing of Personal Science and the Narrative Element in Science Education," *Science Education* 75, no. 6 (1991): 707–22, https://doi.org/10.1002/sce.3730750610.

18. Shawn Wilson, *Research Is Ceremony: Indigenous Research Methods* (Halifax, NS: Fernwood, 2008), 123.

19. Wilson, 17.

20. John Angus Campbell, "The Polemical Mr. Darwin," *Quarterly Journal of Speech* 61, no. 4 (1975): 378, https://doi.org/10.1080/00335637509383301.

21. Bill Nye, *Undeniable: Evolution and the Science of Creation* (New York: St. Martin's Press, 2014), 15.

22. Aïda Diongue Niang, "Conversation Shops in Senegal," Climate Outreach, accessed October 14, 2022, https://climateoutreach.org/case-studies-from-ipcc-authors/senegal/.

23. Intan Suci Nurhati, "Talking Your Audience's Language in Indonesia," Climate Outreach, accessed October 14, 2022, https://climateoutreach.org/case-studies-from-ipcc-authors/indonesia/.

24. Niang, "Conversation Shops in Senegal."

25. These are just a few of the many studies that discuss this unfortunate trend in representations, stereotypes, and assumptions about the identities of scientists. Lillian Campbell, "'MacGyver-Meets-Dr. Ruth': Science Journalism and the Material Positioning of Dr. Carla Pugh," *Women's Studies in Communication* 37, no. 1 (2014): 44–65, https://doi.org/10.1080/07491409.2013.867916; Marcel C. LaFollette, "Eyes on the Stars: Images of Women Scientists in Popular Magazines," *Science, Technology, & Human Values* 13, no. 3–4 (1988): 262–75, https://doi.org/10.1177/016224398801303-407; Yin Kiong Hoh, "Using Biographies of Outstanding Women in Bioengineering to Dispel Biology Teachers' Misperceptions of Engineers," *American Biology Teacher* 71, no. 8 (2009): 458–63, https://doi.org/10.1662/005.071.0804; Jocelyn Steinke et al., "Assessing Media Influences on Middle School–Aged Children's Perceptions of Women in Science Using the Draw-A-Scientist Test (DAST)," *Science Communication* 29, no. 1 (2007): 35–64, https://doi.org/10.1177/1075547007306508.

26. Citizen Science Association, "Citizen Science: Partnering the Public and Professional Scientists," accessed May 24, 2018, http://citizenscience.org/; James Wynn, *Citizen Science in the Digital Age: Rhetoric, Science, and Public Engagement* (Tuscaloosa: University of Alabama Press, 2017), 6.

27. Citizen Science Association, "Citizen Science."

28. Citizen Science Association.

29. National Research Council, *Thinking Evolutionarily: Evolution Education across the Life Sciences—Summary of a Convocation* (Washington, DC: National Academies Press, 2012), 60.

30. Aaron Hess, "'You Don't Play, You Volunteer': Narrative Public Memory Construction in Medal of Honor: Rising Sun," *Critical Studies in Media Communication* 24, no. 4 (2007): 340, https://doi.org/10.1080/07393180701567729.

31. Johnson and Woodbury have compiled lists of climate games. While they are not exhaustive, they are representative of the number and variety of climate-related games that have been developed in recent years. Ellie Johnson, "19 Climate Games That Could Change the Future," *Climate Interactive,* March 9, 2012, https://www.climateinteractive.org/policy-exercises-and-serious-games/19-climate-games-that-could-change-the-future/; Holly Woodbury, "5 Video Games Advocating for Climate Change," *One Green Planet,* 2021, https://www.onegreenplanet.org/environment/5-video-games-advocating-for-climate-change/.

32. Alex Meehan, "'Daybreak Reminds Us If We Can Imagine It, We Can Make It Happen': Pandemic Designer on His Climate Change Awareness Game," *Dicebreaker,* September 29, 2022, https://www.dicebreaker.com/games/climate-crisis/interview/matt-leacock-matteo-menapace-interview.

33. Bez Shahriari, "Some Initial Thoughts after One Play of a Prototype," *BoardGameGeek,* September 21, 2022, https://boardgamegeek.com/thread/2938248/some-initial-thoughts-after-one-play-prototype.

34. Our Climate Voices, accessed September 21, 2020, https://www.ourclimatevoices.org.

35. Climate Stories Project, accessed May 27, 2018, http://www.climatestoriesproject.org/.

36. Voices for Vaccines, accessed July 23, 2018, https://www.voicesforvaccines.org/.

37. Centers for Disease Control, "Pertussis and Whooping Cough," February 14, 2019, https://www.cdc.gov/pertussis/countries/index.html.

38. Centers for Disease Control.

39. van der Linden et al. propose that increasing awareness of the scientific consensus on climate change leads to accepting other climate change beliefs. While some studies have questioned the long-term effectiveness of consensus messaging (e.g., Kahan; Kerr and Wilson), there is evidence that belief in the scientific consensus around climate change positively influences public support. Sander van der Linden et al., "The Scientific Consensus on Climate Change as a Gateway Belief: Experimental Evidence," *PLOS ONE* 10, no. 2 (2015): 1–8, https://doi.org/10.1371/journal.pone.0118489; Sander van der Linden, Anthony Leiserowitz, and Edward Maibach, "The Gateway Belief Model: A Large-Scale Replication," *Journal of Environmental Psychology* 62 (2019): 49–58, https://doi.org/10.1016/j.jenvp.2019.01.009; Dan M. Kahan, "The 'Gateway Belief'

Illusion: Reanalyzing the Results of a Scientific-Consensus Messaging Study," *Journal of Science Communication* 16, no. 5 (2016): 1–20, http://dx.doi.org/10.2139/ssrn.2779661; John Richard Kerr and Marc Stewart Wilson, "Perceptions of Scientific Consensus Do Not Predict Later Beliefs about the Reality of Climate Change: A Test of the Gateway Belief Model Using Cross-Lagged Panel Analysis," *Journal of Environmental Psychology* 59 (2018): 107–10, https://doi.org/10.1016/j.jenvp.2018.08.012.

40. Paul R. Brewer and Jessica McKnight, "'A Statistically Representative Climate Change Debate': Satirical Television News, Scientific Consensus, and Public Perceptions of Global Warming," *Atlantic Journal of Communication* 25, no. 3 (2017): 166–80, https://doi.org/10.1080/15456870.2017.1324453.

41. National Center for Science Education, *Voices for Climate Change Education,* June 5, 2016, https://ncse.ngo/voices-climate-change-education-0.

42. National Center for Science Education, *Voices for Evolution,* 3rd ed. (2008), https://ncse.ngo/voices-evolution-0.

43. Kenneth Burke, *A Grammar of Motives* (Berkeley: University of California Press, 1945), 77.

44. Emma Frances Bloomfield, *Communication Strategies for Engaging Climate Skeptics: Religion and the Environment* (New York: Routledge, 2019), 171.

45. Saffron O'Neill and Sophie Nicholson-Cole, "'Fear Won't Do It': Promoting Positive Engagement with Climate Change Through Visual and Iconic Representations," *Science Communication* 30, no. 3 (2009): 355–79, https://doi.org/10.1177/1075547008329201.

46. P. Sol Hart and Erik C. Nisbet, "Boomerang Effects in Science Communication: How Motivated Reasoning and Identity Cues Amplify Opinion Polarization about Climate Mitigation Policies," *Communication Research* 39, no. 6 (2012): 701–23, https://doi.org/10.1177/0093650211416646.

47. Elizabeth M. DeLoughrey, *Allegories of the Anthropocene* (Durham, NC: Duke University Press, 2019), 20.

48. DeLoughrey, 20; V. Jo Hsu, *Constellating Home: Trans and Queer Asian American Rhetorics* (Columbus: Ohio State University Press, 2022); Aja Y. Martinez, *Counterstory: The Rhetoric and Writing of Critical Race Theory* (Champaign, IL: National Council of Teachers of English, 2020); Linda Tuhiwai Smith, *Decolonizing Methodologies: Research and Indigenous Peoples* (London: Zed Books, 2021).

49. *Mayah's Lot,* Center for Urban Environmental Reform (blog), November 20, 2013, https://cuer.law.cuny.edu/?page_id=1272.

50. Christopher Scott Thomas, "Drawing Environmental Justice: Digital Storytelling and Interventionist Messages in Mayah's Lot" (presentation, National Communication Association 107th Annual Convention, Seattle, WA, November 18–21, 2021).

51. Nadia Ramlagan, "Tribes Plan for Climate Change in Drought-Stricken Nevada," American Association for the Advancement of Science, February 16, 2016, https://www.aaas.org/news/tribes-plan-climate-change-drought-stricken-nevada.

52. Our study found that a story set in the recent past was most effective for self-identified Republican and Independent audiences. This story was also effective for Democrats, but a story told in the present was more effective for them. Sangalang and Bloomfield, "Mother Goose and Mother Nature," 583–604.

53. Esben Bjerggaard Nielsen, "Climate Crisis Made Manifest: The Shift from a Topos of Time to a Topos of Place," in *Topic-Driven Environmental Rhetoric,* ed. Derek G. Ross (New York: Routledge, 2017), 87–105.

54. Naomi Oreskes and Erik M. Conway, "Defeating the Merchants of Doubt," *Nature* 465 (2010): 686–87, https://www.nature.com/articles/465686a.

55. Elisabeth A. Lloyd et al., "Climate Scientists Set the Bar of Proof Too High," *Climatic Change* 165, no. 55 (2021): 2, https://doi.org/10.1007/s10584-021-03061-9.

56. Adam Corner et al., *The Uncertainty Handbook,* Climate Outreach, February 23, 2016, https://climateoutreach.org/reports/uncertainty-handbook/.

57. O'Neill and Nicholson-Cole, "'Fear Won't Do It,'" 375.

58. Mallika Kallingal, "Meet the Illustrators Who Gave the Coronavirus Its Face," *CNN,* April 18, 2020, https://www.cnn.com/2020/04/17/us/coronavirus-cdc-design-trnd/index.html.

59. Kallingal.

60. Cara Giaimo, "The Spiky Blob Seen Around the World," *New York Times,* April 1, 2020, https://www.nytimes.com/2020/04/01/health/coronavirus-illustration-cdc.html.

61. Andy J. King and Allison J. Lazard, "Advancing Visual Health Communication Research to Improve Infodemic Response," *Health Communication* 35, no. 14 (2020): 1724, https://doi.org/10.1080/10410236.2020.1838094.

62. DeLoughrey, *Allegories of the Anthropocene,* 2.

63. DeLoughrey, 34.

64. Anne M. van Valkengoed, Linda Steg, and Goda Perlaviciute, "The Psychological Distance of Climate Change Is Overestimated," *One Earth* 6, no. 4 (2023): 362–91, https://doi.org/10.1016/j.oneear.2023.03.006.

65. Bloomfield, *Communication Strategies for Engaging Climate Skeptics*; for more details on weak and strong sustainability, see Paula Castro, Mehmet ali Uzelgun, and Raquel Bertoldo, "Climate Change Activism between Weak and Strong Environmentalism: Advocating Social Change with Moderate Argumentation Strategies," in *The Social Psychology of Everyday Politics,* ed. Caroline Howarth and Eleni Andreouli (New York: Routledge, 2016), 156–72.

66. Kris De Meyer, Emily Coren, Mark McCaffrey, and Cheryl Slean, "Transforming the Stories We Tell about Climate Change: From 'Issue' to 'Action,'" *Environmental Research Letters* 16 (2021): 1–13, https://doi.org/10.1088/1748-9326/abcd5a.

67. Ramlagan, "Tribes Plan for Climate Change."

68. O'Neill and Nicholson-Cole, "'Fear Won't Do It,'" 371–72, emphasis added.

69. Dylan M. Harris, "Telling Stories about Climate Change," *Professional Geographer* 72, no. 3 (2020): 312, https://doi.org/10.1080/00330124.2019.1686996.

70. DeLoughrey, *Allegories of the Anthropocene,* 107.

71. National Research Council, *Thinking Evolutionarily,* 36.

72. National Research Council, 36.

73. Jeanne Fahnestock, "Accommodating Science: The Rhetorical Life of Scientific Facts," *Written Communication* 15, no. 3 (1998): 330–50, https://doi.org/10.1177/0741088386003003001; Sophie Guy et al., "Comparing the Atmosphere to a Bathtub: Effectiveness of Analogy Reasoning about Accumulation," *Climatic Change* 121 (2013), 579–94, https://doi.org/10.1007/s10584-013-0949-3; Giambattista Vico, *The Art of Rhetoric,* trans. Giorgio A. Pinton and Arthur W. Shippee (Amsterdam: Rodopi, 1996).

74. Ceccarelli, "CRISPR as Agent," 1–13; Emma Frances Bloomfield, "The Rhetoric of Energy Darwinism: Neoliberal Piety and Market Autonomy in Economic Discourse," *Rhetoric Society Quarterly* 49, no. 4 (2019): 320–41, https://doi.org/10.1080/02773945.2019.1634831.

75. Carl Sagan, "Carl Sagan—*Cosmos*—Cosmic Calendar," YouTube, July 22, 2009, science video, 4:57, https://www.youtube.com/watch?v=Ln8UwPd1z20.

76. Sagan, emphasis added.

77. Ed Hawkins, "Show Your Stripes," Institute for Environmental Analytics, accessed October 23, 2022, https://showyourstripes.info/faq.

78. Hawkins.

79. Bobby Bascomb, "'Hadestown' Brings Climate Change to Broadway," *Living on Earth,* July 19, 2019, https://www.loe.org/shows/segments.html?programID=19-P13-00029&segmentID=5.

80. "Road to Hell II" from *Hadestown,* music and lyrics by Anaïs Mitchell, 2016.

81. Margaret Kovach, "Doing Indigenous Methodologies: A Letter to a Research Class," in *The SAGE Handbook of Qualitative Research,* ed. Norman K. Denzin and Yvonna S. Lincoln (Los Angeles: SAGE, 2017), 229.

82. Emma Frances Bloomfield, "Transcorporeal Identification and Strategic Essentialism in Eco-Horror: *mother!*'s Ecofeminist Rhetorical Strategies," *Environmental Communication* 15, no. 3 (2021): 339–52, https://doi.org/10.1080/17524032.2020.1833059.

83. "Deep Time Walk," *Deep Time Walk* (blog), accessed February 1, 2021, https://www.deeptimewalk.org/blog/.

84. Zahraa Saiyed and Paul D. Irwin, "Native American Storytelling toward Symbiosis and Sustainable Design," *Energy Research & Social Science* 31 (2017): 251, https://doi.org/10.1016/j.erss.2017.05.029.

85. Saiyed and Irwin, 251.

86. Saiyed and Irwin, 251.

87. Lauren E. Cagle and Denise Tillery, "Climate Change Research across Disciplines: The Value and Uses of Multidisciplinary Research Reviews for Technical Communication," *Technical Communication Quarterly* 24, no. 2 (2015): 147–63, https://doi.org/10.1080/10572252.2015.1001296.

88. Bloomfield, *Communication Strategies for Engaging Climate Skeptics.*

89. Walter R. Fisher, "Narration as a Human Communication Paradigm: The Case of Public Moral Argument," *Communication Monographs* 51, no. 1 (1984): 14, https://doi.org/10.1080/03637758409390180.

90. Bloomfield, *Communication Strategies for Engaging Climate Skeptics,* 172.

91. The idea that increasing information and discussion of it will close gaps between experts and publics is called the "information deficit model." This model has been proven to have serious flaws in its predictive and explanatory abilities. There are too many studies critiquing the information deficit model to list here, but interested readers may start with Paul G. Bain et al., "Promoting Pro-Environmental Action in Climate Change Deniers," *Nature Climate Change* 2, no. 8 (2012): 600–603, https://doi.org/10.1038/nclimate1532; Emma Frances Bloomfield et al., "The Effects of Establishing Intimacy and Consubstantiality on Group Discussions about Climate Change Solutions," *Science Communication* 42, no. 3 (2020): 369–94, https://doi.org/10.1177/1075547020927017; Lyn M. van Swol et al., "Fostering Climate Change Consensus: The Role of Intimacy in Group Discussions," *Public Understanding of Science* 31, no 1. (2022): 103–18, https://doi.org/10.1177/09636625211020661.

92. Emma Frances Bloomfield, "The Reworking of Evangelical Christian Ecocultural Identity in the Creation Care Movement," in *The Routledge Handbook of Ecocultural Identity,* ed. Tema Milstein and José Castro-Sotomayor (New York: Routledge, 2020), 195–207.

93. Understanding Science, "Science and Religion: Reconcilable Differences," accessed September 28, 2020, https://undsci.berkeley.edu/article/science_religion.

94. Understanding Science.

95. Brandon R. McFadden, "Examining the Gap between Science and Public Opinion about Genetically Modified Food and Global Warming," *PLOS ONE* 11, no. 11 (2016): 1–14, https://doi.org/10.1371/journal.pone.0166140.

96. Doug Cloud, "Communicating Climate Change to Religious and Conservative Audiences," *Reflections* 16 (2016): 57–73, https://reflectionsjournal.net/wp-content/uploads/2016/09/16.1-Website-Entire-Issue.pdf#page=65.

97. Bloomfield, "The Reworking of Evangelical Christian Ecocultural Identity," 195–207.

98. Bloomfield, 195–207.

99. Jennifer A. Reich, *Calling the Shots: Why Parents Reject Vaccines* (New York: NYU Press, 2018).

100. Although there are limited arguments against vaccines from formal religious teachings, and most faiths directly support vaccination, there are circulating arguments, especially in online spaces, that encourage hesitation from religious grounds. For example, stem cells are involved in the process of testing some vaccines, but no stem cells are *in* vaccines. The US Conference of Catholic Bishops encouraged Catholics to opt for the Pfizer or Moderna vaccine over the Johnson & Johnson COVID-19 vaccine for this reason. Additionally, the presence of gelatin, an animal product, in vaccines is viewed by some as nonkosher. Emma Frances Bloomfield and Stephanie S. Willes, "Religious Masking and the Rhetorical Strategies of Digital Anti-Vaccination Churches," *Western Journal of Communication* (2023), https://doi.org/10.1080/10570314.2023.2174384.

101. Bloomfield and Willes.

102. John D. Grabenstein, "What the World's Religions Teach, Applied to Vaccines and Immune Globulins," *Vaccine* 31, no. 16 (2013): 2011–23, https://doi.org/10.1016/j.vaccine.2013.02.026.

103. Donald G. McNeil, Jr., "Religious Objections to the Measles Vaccine? Get the Shots, Faith Leaders Say," *New York Times,* April 26, 2019, https://www.nytimes.com/2019/04/26/health/measles-vaccination-jews-muslims-catholics.html; Nicholas A. Daniels et al., "Effectiveness of Adult Vaccination Programs in Faith-Based Organizations," *Ethnicity and Disease* 17, no. 1 (2007): 15–22, https://www.jstor.org/stable/48667169; UNICEF, *Building Trust in Immunization: Partnering with Religious Leaders and Groups* (2004), https://www.unicef.org/publications/index_20944.html.

104. Wilson, *Research Is Ceremony,* 89.

105. John H. Evans, *Morals Not Knowledge: Recasting the Contemporary U.S. Conflict between Religion and Science* (Oakland: University of California Press, 2018).

106. Bloomfield, "The Rhetoric of Energy Darwinism," 320–41.

107. Bloomfield, *Communication Strategies for Engaging Climate Skeptics.*

108. Bloomfield, "The Rhetoric of Energy Darwinism," 320–41.

109. Doug Cloud, "The Corrupted Scientist Archetype and Its Implications for Climate Change Communication and Public Perceptions of Science," *Environmental Communication* 14, no. 6 (2020): 816–29, https://doi.org/10.1080/17524

032.2020.1741420; Bloomfield, *Communication Strategies for Engaging Climate Skeptics.*

110. Bloomfield, *Communication Strategies for Engaging Climate Skeptics,* 96.

111. Kevin C. Elliott, "Anthropocentric Indirect Arguments for Environmental Protection," *Ethics, Policy & Environment* 17, no. 3 (2014): 244, https://doi.org/10.1080/21550085.2014.955311.

112. Robert G. Eccles, Ioannis Ioannou, and George Serafeim, "The Impact of a Corporate Culture of Sustainability on Corporate Behavior and Performance," *National Bureau of Economic Research* (2012): https://www.nber.org/papers/w17950; Ram Nidumolu, C. K. Prahalad, and M. R. Rangaswami, "Why Sustainability Is Now the Key Driver of Innovation," *Harvard Business Review,* September 1, 2009, https://hbr.org/2009/09/why-sustainability-is-now-the-key-driver-of-innovation.

113. Cristina Pasca Palmer, "Why a Healthy Planet and a Healthy Economy Go Hand-in-Hand," World Economic Forum, January 16, 2019, https://www.weforum.org/agenda/2019/01/save-the-planet-save-the-economy-cristiana-pasca-palmer/.

114. In my first book, I spoke with a climate skeptic named Lucas (pseudonym) who thought that academics receive lots of money for their publications, which he believed incentivizes sloppy work that conforms to the mainstream narrative. In offering myself as a counterexample and walking Lucas through my salary and the publishing process, I corrected misperceptions he had about academic finances. Bloomfield, *Communication Strategies for Engaging Climate Skeptics,* 104–5.

115. Andrew J. Hoffman, "Talking Past Each Other? Cultural Framing of Skeptical and Convinced Logics in the Climate Change Debate," *Organization & Environment* 24, no. 1 (2011): 3–33, https://doi.org/10.1177/1086026611404336.

116. Corynn Oceana Miller and Emma Frances Bloomfield, "'You Can't Be What You Can't See': Analyzing Alexandria Ocasio-Cortez's Environmental Rhetoric," *Journal of Contemporary Rhetoric* 12, no. 1 (2022): 1–16, http://contemporaryrhetoric.com/wp-content/uploads/2022/06/Miller_Bloomfield_12_1_1.pdf.

117. Graham Dixon, Jay Hmielowski, and Yanni Ma, "Improving Climate Change Acceptance among U.S. Conservatives Through Value-Based Message Targeting," *Science Communication* 39, no. 4 (2017): 520–34, https://doi.org/10.1177/1075547017715473; Jack Zhou, "Boomerangs versus Javelins: How Polarization Constrains Communication on Climate Change," *Environmental Politics* 25, no. 5 (2016): 788–811, https://doi.org/10.1080/09644016.2016.1166602; Kerrie L. Unsworth and Kelly S. Fielding, "It's Political: How the Salience of One's Political Identity Changes Climate Change Beliefs and Policy Support," *Global Environmental Change* 27 (2014): 131–37, https://doi.org/10.1016/j.gloenvcha.2014.05.002.

118. Matthew Facciani, "Video: How Did Mask Wearing Become So Politicized?," *The Conversation,* September 9, 2020, http://theconversation.com /video-how-did-mask-wearing-become-so-politicized-144268.

119. Andy Burness, "Bipartisan Science," *Issues in Science and Technology* (blog), September 28, 2015, https://issues.org/bipartisan-science/.

120. Congressional Science Policy Initiative, *Federation of American Scientists* (blog), accessed November 2, 2020, https://fas.org/congressional-science-policy-initiative/; "The Science Coalition," accessed November 2, 2020, https:// www.sciencecoalition.org/.

121. Emma Frances Bloomfield, "Understanding Christians' Climate Views Can Lead to Better Conversations about the Environment," *The Conversation,* August 5, 2019, https://theconversation.com/understanding-christians-climate-views-can-lead-to-better-conversations-about-the-environment-115693.

122. Aristotle, *Rhetoric,* trans. W. Rhys Roberts (Internet Classics Archive), http://classics.mit.edu/Aristotle/rhetoric.1.i.html; Chaïm Perelman, "How Do We Apply Reason to Values?," *Journal of Philosophy* 52, no. 26 (1955): 797–802, https://www.jstor.org/stable/2022489; Fisher, "Narration as a Human Communication Paradigm," 1–22.

123. Smith, *Decolonizing Methodologies;* Daniel Wildcat, *Red Alert!: Saving the Planet with Indigenous Knowledge* (Wheat Ridge, CO: Fulcrum, 2009).

124. Bloomfield et al., "The Effects of Establishing Intimacy and Consubstantiality," 369–94; Hart and Nisbet, "Boomerang Effects in Science Communication," 701–23; Gregory J. Trevors et al., "Identity and Epistemic Emotions during Knowledge Revision: A Potential Account for the Backfire Effect," *Discourse Processes* 53, no. 5–6 (2015): 339–70, https://doi.org/10.1080/0163853X.2015.1136507.

125. Crystal Lee et al., "Viral Visualizations: How Coronavirus Skeptics Use Orthodox Data Practices to Promote Unorthodox Science Online," in *Proceedings of the 2021 CHI Conference on Human Factors in Computing Systems* (New York: Association for Computing Machinery, 2021), 1–18, https://doi .org/10.1145/3411764.3445211.

126. Womanism is a theoretical framework that centers Black women's experiences as a foil to (White) feminism. Audre Lorde, *Sister Outsider: Essays and Speeches* (Toronto: Crossing Press, 1984), 25.

127. Jennie C. Stephens, "Beyond Climate Isolationism: A Necessary Shift for Climate Justice," *Current Climate Change Reports* 8 (2022): 83–90, https://doi .org/10.1007/s40641-022-00186-6.

128. Stephens, 83.

129. Farhana Sultana, "Critical Climate Justice," *Geographical Journal* 188, no. 1 (2022): 118–24, https://doi.org/10.1111/geoj.12417.

130. Farhana Sultana, "The Unbearable Heaviness of Climate Coloniality," *Political Geography* 99 (2022): 1–16, https://doi.org/10.1016/j.polgeo.2022.102638; see

also G. Mitchell Reyes and Kundai Chirindo, "Theorizing Race and Gender in the Anthropocene," *Women's Studies in Communication* 43, no. 4 (2020): 429–42, https://doi.org/10.1080/07491409.2020.1824519.

CONCLUSION

1. Ada, a tabby cat with cinnamon-swirl patterns on her sides, is named after the first computer scientist, Ada Lovelace. Izzy, a tuxedo cat, is named after Sir Isaac Newton, a mathematician famous for developing the laws of gravity.

2. Kundai Chirindo, "Precarious Publics," *Quarterly Journal of Speech* 107, no. 4 (2021): 430–34, https://doi.org/10.1080/00335630.2021.1983192.

3. Aristotle, *Rhetoric,* trans. W. Rhys Roberts (Internet Classics Archive), http://classics.mit.edu/Aristotle/rhetoric.1.i.html.

4. Jeanne Fahnestock, "Rhetorical Citizenship and the Science of Science Communication," *Argumentation* 34, no. 3 (2020): 384, https://doi.org/10.1007/s10503-019-09499-7.

5. Because we do not want stories that have all of their narrative features mapped on the hub spiral, this advice could also be phrased as we should map two to five narrative features on the micro-ring.

6. G. Thomas Goodnight, "Science and Technology Controversy: A Rationale for Inquiry," *Argumentation and Advocacy* 42, no. 1 (2005): 26–29, https://doi.org/ 10.1080/00028533.2005.11821636.

7. Heidi Y. Lawrence, "Fear of the Irreparable: Narratives in Vaccination Rhetoric," *Narrative Inquiry in Bioethics* 6, no. 3 (2016): 205–9, https://doi.org/10.1353/nib.2016.0060.

8. Hoppin's quotation comes from an analysis of vaccine discourse, but it is relevant to genetically modified foods because of its similar content issues regarding health, family decision-making, and children's well-being. Shari Hoppin, "Applying the Narrative Paradigm to the Vaccine Debates," *American Communication Journal* 18, no. 2 (2016): 51.

9. Non-GMO Project, "GMO Facts," accessed April 13, 2023, https://www.nongmoproject.org/gmo-facts/.

10. Non-GMO Project.

11. Ryan McGrady, "The Conversation about GMOs in 2021," Health Discourse Research Initiative, March 9, 2022, https://www.mediaecosystems.org/case-study/the-conversation-about-gmos-in-2021.

12. Danielle Endres, "The Rhetoric of Nuclear Colonialism: Rhetorical Exclusion of American Indian Arguments in the Yucca Mountain Nuclear Waste Siting Decision," *Communication and Critical/Cultural Studies* 6, no. 1 (2009): 39–60, https://doi.org/10.1080/14791420802632103.

13. Mark Z. Jacobson, "7 Reasons Why Nuclear Energy Is Not the Answer to Solve Climate Change," One Earth, September 6, 2022, https://www.oneearth.org/the-7-reasons-why-nuclear-energy-is-not-the-answer-to-solve-climate-change/; Office of Nuclear Energy, "5 Fast Facts about Spent Nuclear Fuel," October 3, 2022, https://www.energy.gov/ne/articles/5-fast-facts-about-spent-nuclear-fuel.

14. Hannah K. Patenaude and Emma Frances Bloomfield, "Topical Analysis of Nuclear Experts' Perceptions of Publics, Nuclear Energy, and Sustainable Futures," *Frontiers in Science and Environmental Communication* 7 (2022): 1–13, https://doi.org/10.3389/fcomm.2022.762101.

15. John E. Kotcher et al., "Does Engagement in Advocacy Hurt the Credibility of Scientists? Results from a Randomized National Survey Experiment," *Environmental Communication* 11, no. 3 (2017): 415–29, https://doi.org/10.1080/17524032.2016.1275736.

16. Jacobson, "7 Reasons Why Nuclear Energy Is Not the Answer"; Ken Silverstein, "Are Fossil Fuel Interests Bankrolling the Anti-Nuclear Energy Movement?" *Forbes,* July 13, 2016, https://www.forbes.com/sites/kensilverstein/2016/07/13/are-fossil-fuel-interests-bankrolling-the-anti-nuclear-energy-movement/?sh=4dfeef967453; Justin Rowlatt, "Nuclear Power: Are We Too Anxious about the Risks of Radiation?" *BBC News,* September 27, 2020, https://www.bbc.com/news/science-environment-54211450.

17. Retrieved from Steve Shwartz, "Are Self-Driving Cars Really Safer than Human Drivers?" *The Gradient,* June 13, 2021, https://thegradient.pub/are-self-driving-cars-really-safer-than-human-drivers/.

18. Calum Mckinney, "Study: Self-Driving Cars 'Learn' How to Make Moral Decisions on the Road," *Study Finds,* August 4, 2017, https://studyfinds.org/self-driving-cars-morality-ai/.

19. Shwartz, "Are Self-Driving Cars Really Safer than Human Drivers?"

20. Tatiania Perry, "Study: Majority of Drivers Don't Feel Comfortable in Autonomous Vehicles," *Newsweek,* March 7, 2022, https://www.newsweek.com/study-majority-drivers-dont-feel-comfortable-autonomous-vehicles-1685089; Alex Mitchell, "Are We Ready for Self-Driving Cars?" *World Economic Forum,* November 24, 2015, https://www.weforum.org/agenda/2015/11/are-we-ready-for-self-driving-cars/.

21. Brooke Niemeyer, "More than 3 in 4 Americans Feel Less Safe in Self-Driving Cars, Survey Finds," *PR News Wire,* September 14, 2022, https://www.prnewswire.com/news-releases/more-than-3-in-4-americans-feel-less-safe-in-self-driving-cars-survey-finds-301623530.html.

22. Sabrina Ortiz, "What Is ChatGPT and Why Does It Matter? Here's What You Need to Know," *ZDNET,* April 18, 2023, https://www.zdnet.com/article/what-is-chatgpt-and-why-does-it-matter-heres-everything-you-need-to-know/; Miley Shen, "Understanding Chat GPT: What It Is and How to Use It," *Awesome Screenshot,* March 29, 2023, https://www.awesomescreenshot.com/blog

/knowledge/what-is-chat-gpt; Yvette Mucharraz y Cano, Francesco Venuity, and Ricardo Herrera Martinez, "ChatGPT and AI Text Generators: Should Academia Adapt or Resist?" *Harvard Business Publishing,* January 31, 2023, https://hbsp.harvard.edu/inspiring-minds/chatgpt-and-ai-text-generators-should-academia-adapt-or-resist.

23. Leonardo De Cosmo, "Google Engineer Claims AI Chatbot Is Sentient: Why That Matters," *Scientific American,* July 12, 2022, https://www.scientificamerican.com/article/google-engineer-claims-ai-chatbot-is-sentient-why-that-matters/.

24. Emma Frances Bloomfield, "The Rhetoric of Energy Darwinism: Neoliberal Piety and Market Autonomy in Economic Discourse," *Rhetoric Society Quarterly* 49, no. 4 (2019): 320–41, https://doi.org/10.1080/02773945.2019.1634831; Tema Milstein, "Nature Identification: The Power of Pointing and Naming," *Environmental Communication* 5, no. 1 (2011): 3–24, https://doi.org/10.1080/17524032.2010.535836.

25. Laurie Clarke, "Alarmed Tech Leaders Call for AI Research Pause," *Science,* April 11, 2023, https://www.science.org/content/article/alarmed-tech-leaders-call-ai-research-pause.

26. Clarke.

27. Brian McCammack, "Hot Damned America: Evangelicalism and the Climate Change Policy Debate," *American Quarterly* 59, no. 3 (2007): 645–68, https://doi.org/10.1353/aq.2007.0065; Emma Bloomfield, "Rhetorical Strategies in Contemporary Responses to Science and Modernity: Legitimizing Religion in Human Origins and Climate Change Controversies" (PhD diss., University of Southern California, 2016), http://search.proquest.com/docview/2070186953/?pq-origsite=primo.

28. Lisa Martine Jenkins, "How Concern Over Climate Change Correlates with Coronavirus Responses," *Morning Consult,* April 27, 2020, https://morningconsult.com/2020/04/27/climate-change-concern-coronavirus-response-polling/.

29. For example, see Emma Frances Bloomfield et al., "The Effects of Establishing Intimacy and Consubstantiality on Group Discussions about Climate Change Solutions," *Science Communication* 42, no. 3 (2020): 369–94, https://doi.org/10.1177/1075547020927017; Caitlin Drummond and Baruch Fischhoff, "Individuals with Greater Science Literacy and Education Have More Polarized Beliefs on Controversial Science Topics," *Proceedings of the National Academy of Sciences* 114, no. 36 (2017): 9587–92, https://doi.org/10.1073/pnas.1704882114; Lawrence C. Hamilton, Joel Hartter, and Kei Saito, "Trust in Scientists on Climate Change and Vaccines," *SAGE Open* 5, no. 3 (2015), https://doi.org/10.1177/2158244015602752; John R. Kerr and Marc S. Wilson, "Right-Wing Authoritarianism and Social Dominance Orientation Predict Rejection of Science and Scientists," *Group Processes & Intergroup Relations* 24, no. 4 (2021):

550–67, https://doi.org/10.1177/1368430221992126; Bastiaan T. Rutjens, Robbie M. Sutton, and Romy van der Lee, "Not All Skepticism Is Equal: Exploring the Ideological Antecedents of Science Acceptance and Rejection," *Personality and Social Psychology Bulletin* 44, no. 3 (2018): 384–405, https://doi.org/10.1177/0146167217741314.

30. For example, in a previous study, Sangalang and I modified story elements based on temporality, morality, and realism. Angeline Sangalang and Emma Frances Bloomfield, "Mother Goose and Mother Nature: Designing Stories to Communicate Information about Climate Change," *Communication Studies* 69, no. 5 (2018): 583–604, https://doi.org/ 10.1080/10510974.2018.1489872.

31. Visit Washington DC, "6 Reasons Why You Must Visit the New Fossil Hall at the National Museum of Natural History," April 13, 2023, https://washington.org/visit-dc/deep-time-exhibit-natural-history-museum#.

32. The term *Te Taiao* means "nature" in Māori and is used in the expression "Ko au te taiao, te taiao ko au [I am nature, nature is me]." Museum of New Zealand, "Te Taiao | Nature," accessed April 13, 2023, https://www.tepapa.govt.nz/visit/exhibitions/te-taiao-nature.

33. Lili Pâquet, "A Rhetoric of Walking and Reading: Immersion in Environmental Ambient Literature," *Rhetoric Society Quarterly* 50, no. 4 (2020): 269, https://doi.org/10.1080/02773945.2020.1748216.

34. Paula Castro, Mehmet ali Uzelgun, and Raquel Bertoldo, "Climate Change Activism between Weak and Strong Environmentalism: Advocating Social Change with Moderate Argumentation Strategies," in *The Social Psychology of Everyday Politics*, ed. Caroline Howarth and Eleni Andreouli (New York: Routledge, 2016), 156–72.

35. Madison P. Jones, "Portfolio: DWELL Lab," accessed February 1, 2021, http://madisonpjones.com/pages/dwell.html.

36. DWELL Lab, "AR Map Tour of Block Island," accessed October 17, 2022, https://web.uri.edu/dwell/projects/manissean-tour-map/.

37. Julie Collins Bates, "Local Expertise, Global Effects: Amplifying Activist Arguments for Climate Change Action," *Enculturation: A Journal of Rhetoric, Writing, and Culture* (2020), http://enculturation.net/Local_Expertise_Global_Effects.

38. For example, see Lauren E. Cagle, "Climate Change and the Virtue of Civility: Cultivating Productive Deliberation around Public Scientific Controversy," *Rhetoric Review* 37, no. 4 (2018): 370–79, https://doi.org/10.1080/07350198.2018.1497882; Emma Frances Bloomfield and Denise Tillery, "The Circulation of Climate Change Denial Online: Rhetorical and Networking Strategies on Facebook," *Environmental Communication* 13, no. 1 (2019): 23–34, https://doi.org/10.1080/17524032.2018.1527378.

39. Deplatforming is a controversial strategy for moderating digital spaces. Some argue that deplatforming is a violation of free speech. These claims are

rooted in the notion that people should be free to say whatever they want whenever and wherever they want, when the First Amendment only protects from government censorship, and not necessarily from businesses or social media platforms. Other opponents to deplatforming argue that such actions inadvertently raise attention to a subject. Proponents of deplatforming argue that it is effective in regulating toxic users and content. For example, information and computer scientists Jhaver and colleagues found that deplatforming provocateurs on Twitter decreased conversation about them by 91.77 percent on average and reduced previous followers' content toxicity by 5.84 percent (measured by authors through a toxicity score). Shagun Jhaver et al., "Evaluating the Effectiveness of Deplatforming as a Moderation Strategy on Twitter," *Proceedings of the ACM on Human-Computer Interaction* 5 (2021): 1–30, https://doi.org/10.1145/3479525.

40. Gabriela Czarnek, Małgorzata Kossowska, and Paulina Szwed, "Right-Wing Ideology Reduces the Effects of Education on Climate Change Beliefs in More Developed Countries," *Nature Climate Change* 11 (2020): 9–13, https://doi.org/10.1038/s41558-020-00930-6; see also Roger Few et al., "Culture as a Mediator of Climate Change Adaptation: Neither Static nor Unidirectional," *WIREs Climate Change* 12, no. 1 (2021): 1–8, https://doi.org/10.1002/wcc.687; Melbourne S. Cummings and Judi Moore Latta, "When They Honor the Voice: Centering African American Women's Call Stories," *Journal of Black Studies* 40, no. 4 (2010): 666–82, https://doi.org/10.1177/0021934708318666; Gianna M. Savoie, "I Am Ocean: Expanding the Narrative of Ocean Science through Inclusive Storytelling," *Frontiers in Communication* 5 (2020), https://doi.org/10.3389/fcomm.2020.577913.

41. For poignant examples of labor and exclusion in academia, see Ahmed and Reyes. Sara Ahmed, *Complaint!* (Durham, NC: Duke University Press, 2021); Victoria Reyes, *Academic Outsider: Stories of Exclusion and Hope* (Redwood City, CA: Stanford University Press, 2022).

42. Nina Wallerstein and Bonnie Duran, "Using Community-Based Participatory Research to Address Health Disparities," *Health Promotion Practice* 7, no. 3 (2006): 312, https://doi.org/10.1177/1524839906289376.

43. Jessica Hernandez, *Fresh Banana Leaves: Healing Indigenous Landscapes through Indigenous Science* (Berkeley, CA: North Atlantic Books, 2022), 87, 108.

44. Tuck describes the harm of "damage-centered" research on Indigenous communities and Simpson points out the value of incorporating community input into the research process. Eve Tuck, "Suspending Damage: A Letter to Communities," *Harvard Educational Review* 79, no. 3 (2009): 409–28, https://doi.org/10.17763/haer.79.3.n0016675661t3n15; Audra Simpson, "On Ethnographic Refusal: Indigeneity, 'Voice' and Colonial Citizenship," *Junctures: The Journal for Thematic Dialogue* 9 (2007): 67–80, https://junctures.org/junctures/index.php/junctures/article/view/66/60.

45. Hernandez, *Fresh Banana Leaves.*

46. Ali Ersen Erol, "Coherence Co-constructed: Using Coherence for Analysis and Transformation of Social Conflicts," *Narrative and Conflict: Explorations in Theory and Practice* 2, no. 1 (2015): 65, https://doi.org/10.13021/G81019.

47. Nidhi Subbaraman, "How #BlackintheIvory Put a Spotlight on Racism in Academia," *Nature* 582, no. 7812 (2020): 327, https://doi.org/10.1038/d41586-020-01741-7.

48. Naomi Oreskes and Erik M. Conway, "Defeating the Merchants of Doubt," *Nature* 465 (2010): 687, https://www.nature.com/articles/465686a.

49. Oreskes and Conway, 687.

50. Ahmed, *Complaint!*

51. Ayana Elizabeth Johnson and Katharine K. Wilkinson, eds., *All We Can Save: Truth, Courage, and Solutions for the Climate Crisis* (New York: One World, 2020).

52. Nakate is a Ugandan climate activist who started the Rise Up Movement, which highlights the climate stories of people in Africa. Basu is an Emirati-Canadian activist who started the Green Hope Foundation to empower global youth around climate change. "1MILLION Activist Stories," Rise Up Movement Africa, accessed October 18, 2022, https://www.riseupmovementafrica.org/; "Kehkashan Basu—Green Hope Foundation," One Girl, January 12, 2019, https://1girl.ca/kekashan-basu/.

53. Zahraa Saiyed and Paul D. Irwin, "Native American Storytelling toward Symbiosis and Sustainable Design," *Energy Research & Social Science* 31 (2017): 250, https://doi.org/10.1016/j.erss.2017.05.029.

54. Marco Marani et al., "Intensity and Frequency of Extreme Novel Epidemics," *Proceedings of the National Academy of Sciences* 118, no. 35 (2021), https://doi.org/10.1073/pnas.2105482118; Rebecca M. Rice and Jody L. S. Jahn, "Disaster Resilience as Communication Practice: Remembering and Forgetting Lessons from Past Disasters through Practices That Prepare for the Next One," *Journal of Applied Communication Research* 48, no. 1 (2020): 136–55, https://doi.org/10.1080/00909882.2019.1704830.

55. Walter R. Fisher, "Narration as a Human Communication Paradigm: The Case of Public Moral Argument," *Communication Monographs* 51, no. 1 (1984): 1–22, https://doi.org/10.1080/03637758409390180.

56. Michelle Bastian, "Haraway's Lost Cyborg and the Possibilities of Transversalism," *Signs: Journal of Women in Culture and Society* 31, no. 4 (2006): 1030, https://doi.org/10.1086/500597.

57. Fisher, "Narration as a Human Communication Paradigm," 13.

Index

Note: Page numbers in *italics* refer to illustrative matter.

Milton Keynes UK
Ingram Content Group UK Ltd.
UKHW020733060524
442199UK00003B/30